Sanitary Techniques
in Foodservice

Sanitary Techniques in Foodservice

Second Edition

Karla Longrée, Ph.D.
Professor Emeritus, Nutritional Sciences
New York State College of Human Ecology
Cornell University,
Ithaca, New York

Gertrude G. Blaker, Ph.D.
Professor of Food Science and Nutrition
Colorado State University,
Fort Collins, Colorado

1807 1982

John Wiley & Sons
New York • Chichester • Brisbane • Toronto • Singapore

Cover photo-collage by Deborah H. Payne

Library of Congress Cataloging in Publication Data:

Longrée, Karla, 1905-
Sanitary techniques in foodservice.

Includes bibliographical references and index.
1. Food service—Sanitation.   2.   Food handling.
I.   Blaker, Gertrude G., joint author.   II.   Title.
TX911.3.S3L66   1982   363.7'296   81-3047.

ISBN 0-02-371550-2

Preface

Foodservice systems are undergoing constant change, primarily because of the increasing use of foods that are partially or fully prepared away from the premises. Any change in foodservice systems that results from changing the food supply, equipment, and personnel may result in a change in the microbiological problems that confront the system. Therefore, foodservice managers must be professionally equipped to deal with these problems and to teach those under their supervision the appropriate sanitary techniques necessary for their solution.

In preparing the second edition of *Sanitary Techniques in Foodservice,* we have updated information where necessary and have continued to emphasize the theory and application of good sanitation needed by students preparing for positions in the foodservice industry—whether they are being trained on the junior college level, in in-service training programs, or within the framework of other programs that have good sanitation as their goal.

The book is written to serve as:

1. A text for teachers and students of junior colleges, community colleges, technical and trade institutes, special culinary schools, and similar institutions.
2. A teaching aid for teachers of vocational high schools and special training schools for adults; for instructors developing materials for educating food handlers; for teachers in human resource programs; and those involved in similar training programs.
3. A reference book for practitioners to help solve problems and plan in-service training programs.

In this country we have an expanding field of employment for middle- and first-line management personnel in the foodservice industry. We have an increasing and unmet need for administrative dietitians, food-production managers, dietetic technicians, dietetic assistants, foodservice supervisors in hospitals, school foodservice managers, consulting dietitians, dietary supervisors and cook managers in nursing homes, and similar technical positions in other types of establishments. Technical courses in foodservice management are offered in the schools previously listed and these schools attract a sizable number of people who do not plan to attend a four-year course of study in a university.

Sanitary Techniques in Foodservice consists of four parts: Part One, Food Sanitation and Microbiology; Part Two, Food Spoilage and Foodborne Illnesses; Part Three, Sanitary Practice; and Part Four, Education and Training in Sanitation of Foodservice Personnel. Accepted facts are stated but controversial literature is not discussed. Directions are given about purchasing a sound food supply, the sanitary techniques of preparing, holding, and serving food, and the sanitary care of the physical plant, equipment, and utensils. Certain scientific background information is presented under *Readings* for the benefit of teachers and those students who desire to study in greater depth. Study Questions and suggested Readings follow each part.

We hope that the future food handler will become convinced that the public health approach to food sanitation is one of preventing hazards from arising, and will realize that the food handler plays a very important role in the business of serving wholesome meals to the public.

KARLA LONGREE
GERTRUDE G. BLAKER

Contents

List of Tables

List of Procedures

PART ONE

Food Sanitation and Microbiology

SECTION A

The Meaning
of Food Sanitation

In the United States many cases of foodborne illness occur each year. In general, more outbreaks occur in foodservice establishment than in the home, although the reverse may be true in some years. This is serious because eating away from home is very popular, partly because of necessity, partly because of the pleasure derived from eating out. The number of foodservice establishments increases constantly and it is reasonable to assume that approximately one-third to one-half of all meals consumed are purchased and eaten away from home. Outbreaks of food poisoning are usually very dramatic when they occur in a public eating place because of the large number of people apt to be afflicted. Food poisoning in the home often goes unnoticed by the press. In fact, many outbreaks are never recognized as food poisoning and thus go unreported.

A food poisoning outbreak can be defined as *the occurrence of foodborne illness involving two or more people who have partaken of the same food item, with that item identified as the cause of the illness by appropriate scientific analysis.*

A report of such an outbreak may read like this: "More than two hundred children became ill after eating chicken-salad sandwiches; a number of victims had to be hospitalized. . . . The chicken salad had been prepared from warm ingredients and had been left out in the kitchen for several hours. . . ." Or it may read this way: "A festive banquet ended sadly when more than half the guests became ill with violent nausea, vomiting, and diarrhea during the night. . . . Shrimp cocktail was determined to be the offending food. The frozen cooked shrimp had been defrosted overnight in warm water and had remained for an unknown number of hours at temperatures at which rapid bacterial growth was possible. . . . The shrimp were chilled before they were served, but the chilling would not eliminate the bacteria. . . ." As you may have noticed, bacteria caused these foodborne illnesses.

Foodborne illnesses occur even though the food-processing industry, aided by government health authorities, endeavors to do a good job in providing the foodservice industry with a sound food supply. Partially prepared "convenience foods" and fully prepared "ready foods" are available to the foodservice manager in ever-increasing numbers and variety. These foods have certainly changed the activities of most foodservice establishments in this country since those that prepare all menu items "from scratch" are becoming a rarity. Despite the extensive use of convenience and ready-food items, a report of the Center of Disease Control* showed that a total of 99 foodborne disease outbreaks, only 8.7 percent were caused by mishandling food in food-processing plants. The majority of outbreaks (47.8 percent) occurred in foodservice establishments; the remaining outbreaks were either traced to homes (26.1 percent) or were classified as unknown or unspecified (17.4 percent) regarding their origin. In view of these data it must be concluded that managers of foodservice establishments and their staffs need to improve their foodhandling techniques if this situation is to change. A functional foodservice sanitation program should be carefully developed to serve a particular facility and its particular needs. The program should take into account the various danger points at which the food might be subjected to mishandling, such as chances for contamination and holding times and temperatures favorable to progressive multiplication of the contaminants. (See Longrée, K., Chapter XIV*).

Preparing and serving wholesome food to the public is a very important obligation of the foodservice industry. It is an obligation that can only be fulfilled if everyone in every establishment understands what sanitation is, appreciates its importance, and practices it in whatever task is performed.

What is the meaning of sanitation in foodservice? Food sanitation pertains to both the cleanliness and the microbiological, chemical, and physical wholesomeness of food—food that is appealing to the eye and the palate, and that will not make us ill. Practicing sanitation means applying sanitary measures at every stage of the operation—purchasing, receiving, storing, preparing, and serving—for the sake of cleanliness and for the sake of protecting the health of the public served.

Microbiological hazards involve contamination of food with pathogenic microorganisms, such as bacteria, viruses, fungi, protozoa, and parasites. Chemical hazards are concerned with the contamination of food by chemicals that may get into the food accidentally or through purposeful addition. Physical hazards result from contamination of food with injurious particles such as glass chips and metal, wood or bone splinters.

Sanitary practice is concerned with the purchase of a sound food supply and its sanitary storage; with the adequacy of the physical plant and its

*Annual Summary 1976, issued 1977; Atlanta, Ga.
†Also, see the Readings at the end of Part One.

maintenance regarding repairs and cleanliness; with the adequacy and cleanliness of storage facilities; equipment, and utensils; with sanitary dishwashing operations; with the good health, good personal hygiene, and good working habits of the food handler; with the sanitary manipulation of food and effective time-temperature control throughout preparation and service; and finally, with the education of foodservice employees in the various aspects of sanitation in a foodservice operation.

It has been aptly stated by the National Sanitation Foundation,* (a nonprofit, noncommercial organization that seeks and provides solutions to problems involving cleanliness in foodservice), that:

> Sanitation is a way of life. It is the quality of living that is expressed in the clean home, the clean farm, the clean business and industry, the clean neighborhood, the clean community. Being a way of life, it must come from within the people; it is nourished by knowledge and grows as an obligation and an ideal in human relations.

For all who are in the foodservice business, whether they are managers, supervisors, or workers, sanitation can become a way of life through understanding and knowledge. Everyone needs to become aware of and understand the reasons for the evil and dangerous consequences that lack of sanitation creates, and then go ahead and take the second step, the positive approach of action, the unrelenting practice of sanitation, around the clock, every day of the week, every week of the year.

Foodservice operators and food handlers have control over the health of millions of people. It is largely their responsibility whether the food they serve to the public is wholesome. Therefore, foodservice operators and food handlers are part of "public health." It is hoped that in the near future, every person in charge of foodservice operation and every food handler will be a trained person, a professional. Some formal training in basic principles of foodservice should be required of everyone who handles food, and sanitation should be an important part of such training. The training should be terminated by an examination which the trainee must successfully pass before being allowed to operate or manage a foodservice, or to handle food served to the public.

*Ann Arbor, Michigan.

SECTION B

Basic Facts About Microorganisms

The majority of foodborne illnesses are caused by bacteria. A familiarity with some basic facts about bacteria, as well as with some other pathogenic microorganisms is absolutely essential for understanding the meaning of sanitation in foodservice and promoting the techniques for achieving high sanitary standards. Sanitary measures are largely concerned with removal or other effective control of microorganisms in food and everything that makes contact with food.

Microorganisms are microscopically small, living organisms that take up nourishment, give up waste products, and multiply. Some are our friend, some our enemy, and some neither.

Microorganisms are important in connection with food sanitation, because certain disease-producing kinds may be transmitted through food, and other kinds may multiply in food to high numbers and cause so-called food poisoning. Most microorganisms causing foodborne illnesses are bacteria; but viruses, protozoa, and trichinae are other microorganisms that can cause foodborne illnesses. Even certain molds may render food unwholesome.

Microorganisms, especially bacteria and molds, are easily spread and found almost everywhere. We should know what they look like, how they grow, and how they may be killed. We will understand foodborne illnesses better if we understand these facts; and we will understand that we can do a great deal about preventing these incidences by keeping our food from becoming contaminated; by avoiding multiplication of contamininants in the food; by killing the contaminants by proper heating of the food; by removing contaminants from tables, chopping boards, and other equipment and utensils through proper cleaning; and by using other sanitary practices described in this book.

BACTERIA

Bacteria and People

Some kinds work for us, some against us, and some do neither.

Many kinds of bacteria are beneficial; others are neither beneficial nor harmful; still others are undesirable but not harmful; and some are harmful because they spoil our food or are illness-producing or pathogenic to humans.

Beneficial bacteria aid in the manufacture of food items such as cheese, buttermilk, sauerkraut, and pickles.

Benign bacteria are neither beneficial nor harmful. The majority of bacteria belong in this group.

Spoilage bacteria are capable of decomposing our food supply and causing economic loss on all levels—the world, the nation, and the individual wholesaler, retailer, and foodservice operator.

Pathogenic bacteria are capable of causing illness. This latter group is very important to us from a public-health point of view, and therefore requires our attention throughout this book.

Certain pathogenic bacteria are simply transmitted by way of food without growing in that food. These pathogens usually originate from food handlers who have communicable diseases or are seemingly healthy, yet carry pathogens in their system. In this case, food serves as a mere vehicle of transmission—in principle not very different from other vehicles of disease transmission, such as door knobs, money, and other objects likely to be touched by sick persons.

Microorganisms causing food poisoning, however, are different in that they have to first multiply to high numbers in food before they are able to poison those individuals who eat that food. Some food-poisoning microorganisms cause infections, others intoxications. We will discuss this in greater detail later.

Size and Shape

Bacteria are single-celled, thin-walled living organisms that take in nourishment, produce waste, and multiply. They may be round cocci or elongated rods, or they may be commalike or corkscrewlike spirilla. The round cocci may cling together in various numbers and patterns: diplococci are pairs of cells, staphylococci form grapelike clusters, and streptococci form chains. Some rods, too, may cling together in chains (Figure 1). Bacteria are microscopically small and their size varies with species, age, and environmental conditions. The majority of bacteria capable of causing

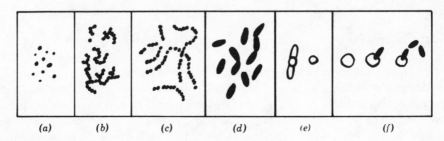

Figure 1. Some bacterial shapes. *(a)* Single cocci. *(b)* Staphylococci. *(c)* Streptococci. *(d)* Rods. *(e)* Spores. *(f)* Germinating spore.

foodborne illness measure approximately $0.5-2 \times 2.0-10.0$ microns, with one micron equivalent to the millionth part of one meter, 1/1000 mm, or 1/25,400 inch.

Motility

Some bacterial forms are equipped with whiplike appendages, the flagella, by which some locomotion is possible. However, the distances that may be travelled are very short. When the flagella originate at one or both ends, this terminal position is referred to as being polar. When the flagella are distributed all over the cell that position is called "peritrichous." All forms of spirilla have flagella.

In spite of the locomotion possible through the aid of flagella, the major means of transportation of bacteria from one place to another are: moving air; person to person contact; contaminated water, food, and food-contact surfaces; insects, rodents, and other animals; and sewage. Rods with flagella are depicted in Figure 2.

Spores

The majority of bacteria possess a fragile and thin wall around the cell; they are called vegetative cells. Some rod-shaped bacteria are able to produce spores that are thick-walled and tough. Spores are able to withstand cold, heat, drought, and other adverse conditions that the thin-walled vegetative cells cannot endure.

When conditions are favorable, spores are able to germinate (Figure 1*f*). The new growth subsequently gives rise to a new crop of thin-walled, vegetative cells capable of dividing further through fission.

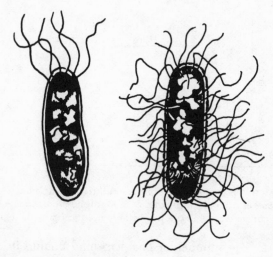

Figure 2. Rods with flagella.

Reproduction

When thin-walled bacterial cells multiply, they simply split in two. This process repeats itself many times, as long as conditions are favorable. If nourishment is of the right kind and plentiful, if temperature conditions are favorable, and if there are no growth-inhibiting substances present, reproduction can proceed very rapidly.

On the other hand, spores must germinate before they can multiply.

The Growth Curve

When bacterial cells multiply by splitting, they pass through various stages. When scientists study the multiplication of bacteria, they count the number of cells over a certain period of time and plot the results on graph paper. Figure 3 shows a typical growth curve. When bacteria multiply they may cluster together, forming colonies. These are best seen when bacteria are cultivated on solid media in the laboratory. When dispersed in a clear liquid medium or a clear food such as consommé, bacterial growth may at times be visible as cloudiness. However, one cannot rely upon the absence of contaminants in food by one's failure to detect them; a bacteriological analysis would have to be made.

The time span over which the growth curve stretches and the highest number of cells produced depend on many factors. Under conditions favorable for growth, a high number of cells is reached in a short time—in fact, in

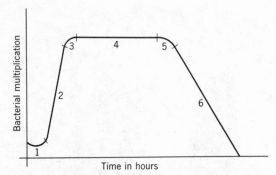

Figure 3. Six stages in the growth curve of bacteria. (1) The lag phase—no multiplication occurs. (2) The phase of rapid multiplication. (3) Multiplication becomes much slower. (4) Growth and death cancel out. (5) Death proceeds faster than growth. (6) Death phase.

just a few hours. The number of bacteria may double in as little as 15 minutes. Here is an example: if we had only one single cell and it divided once in 15 minutes, and every young cell divided every 15 minutes thereafter, this first cell would multiply into approximately 1,000,000 (one million) cells in as little as 5 hours. If for some reason the division of cells is slowed down and takes twice as long, or 30 minutes, production of the same one million cells takes as long as 10 hours.

We must keep this important fact before us, because it is applicable to conditions in food service. We must learn under what circumstances of food preparation bacteria are likely to multiply rapidly and what we can do to prevent this.

Toxin Production

While actively multiplying in food, some bacteria are able to form poisonous toxins that they release into the food. Some of these toxins are so powerful that the person eating the poisoned food can become seriously ill and even die. An example of a deadly poison is the dreaded botulinum toxin. Other toxins are violent in their immediate effect, but their victims usually recover. Again, others are quite mild; to become ill, a person would have to eat a great deal of the poisonous food or to be rather sensitive to the toxin.

We will learn more about toxins in connection with food poisoning (Part Two, Section C).

Conditions for Growth and Toxin Production

In order to multiply, bacteria must have food, moisture, and warmth. Some bacteria must have air and others do better without air; for still others, it does not seem to matter one way or the other.

Food. The various kinds of bacteria have different food needs. However, the forms capable of causing foodborne illnesses prefer foods of a proteinaceous nature, such as milk, meat, poultry, eggs, and seafood.

Moisture. Bacteria require moisture for growth. Dry foods such as sugar, flour, dry cereal, dry rice, cookies, biscuits, and other dry baked products are not good media for bacterial multiplication. Unfortunately, a great number of the prepared menu items that we serve contain sufficient moisture for bacterial growth.

The moisture of food can be expressed as *available moisture* rather than total moisture, since part of the total moisture may be bound in such a way that it is unavailable to microorganisms for their growth. Available moisture is expressed by the term "water activity" and carries the symbol a_w. Water activity is calculated by dividing the vapor pressure of a solution by the vapor pressure of a solvent. Thus, pure water carries a value of 1.00. The addition of salt and sugar to water lowers its water activity, making a portion of the total water less available to microorganisms for their growth. This is the reason for using substances like salt and sugar in the preservation of some foods.

Warmth. The majority of bacteria thrive best at moderately warm temperatures, from 60 to 120 F (15.6 to 49 C); they are called mesophilic bacteria. Disease-producing bacteria, or pathogens, thrive best near body temperature. In general, multiplication is slowed down as temperatures become warmer or cooler than the optimum range; it ceases near or above 140 F (60 C) and above, and at temperatures 45 F (7.2 C) and below. These temperatures play an important role in the application of time-temperature control as a method of producing microbiologically safe food and holding and serving it under safe conditions.

Some bacteria thrive best at high temperatures, 110 to 130 F (43.4 to 54.4 C). These are the heat-loving, or thermophilic, bacteria. Among these are the bacilli causing flat-sour spoilage of canned goods stored at unduly high temperatures. Other bacteria are able to grow well in a cool environment— in fact, at refrigerator temperatures or below. They are the cold-loving, or psychrophilic, bacteria. They are frequently responsible for spoiling food stored too long in the refrigerator.

Bacteria do not multiply in solidly frozen foods, but freezing does not kill all bacteria, either. About one half of the bacteria survive in freezing of foods and the average periods of freezer storage. The survivors may again multiply when the food has been defrosted and has warmed up to 45 F (7.2 C) and above.

For all practical purposes in connection with food preparation, bacterial growth must be expected to take place within the temperature range of 45 and 140 F (7.2 and 60 C). From a public health point of view, this range is called the **DANGER ZONE.** From research data we know that both time and temperature must be controlled. Food should not remain within the danger zone for more than 4 hours; and should not be within the zone of optimum growth (60 to 120 F; 15.6 to 49 C) for more than 2 hours.

Oxygen Requirement.   The requirement for oxygen varies with the kind of bacteria. Some need free oxygen for growth; these are called aerobics or air loving. Others are sensitive to free oxygen and are referred to as anaerobics; they are capable of multiplication in sealed jars and cans. Many bacteria are able to adapt to the presence or absence of free oxygen; they are referred to as the facultative forms of bacteria. Among the facultative group are the majority of the bacteria capable of causing foodborne illnesses.

Acidity.   Most bacteria grow best in media or foods that are neutral or only slightly acid. Most proteinaceous foods such as meat, poultry, eggs, milk, and seafood are nonacid. Most vegetables are slightly acid. But fruits, tomatoes, pickles, and items rendered acid through the addition of vinegar, lemon juice, or other acid ingredients are acid, and are therefore least liked by bacteria as media for their multiplication.

The acidity, neutrality, and alkalinity of a substance is expressed as its concentration of hydrogen ions, or pH. The pH scale extends from 0 (very acid) to 14 (very alkaline). A pH value of 7 signifies neutrality.

The bacteria capable of causing foodborne illnesses will cease to multiply near a pH of 4.6.

Examples of the pH values of some common foods are:

Approximate pH Value	*Food*
2.2−2.6	Lemons
2.9	Vinegar
3.9	Pears
4.0−4.6	Tomatoes
5.0−5.6	Many vegetables
6.2−6.3	Peas, corn
6.4−6.5	Egg yolk
6.5−6.6	Milk
7.0 and lower	Meat
8.0 and higher	Egg white

On the strength of facts known about the growth requirements of pathogens capable of causing foodborne illness, the Food and Drug Administration (U.S. Department of Health, Education and Welfare, Public Health Service) in its *Food Service Manual** has given the following definition for foods that are potentially hazardous because of their ability to support the growth of food-poisoning microorganisms:

"Potentially hazardous food" means any food that consists in whole or in part of milk or milk products, eggs, meat, poultry, fish, shellfish, edible crustacea, or other ingredients, including synthetic ingredients, in a form capable of supporting rapid and progressive growth of infectious or toxigenic microorganisms. The term does not include clean, whole, uncracked, odor-free shell eggs or foods which have a pH level of 4.6 or below or a water activity (a_w) value of 0.85 or less.

Death of Bacteria

Bacteria share with other living beings the fate that sometime they must die. Bacteria die of old age, or because microbial waste products poison them, or because other microorganisms that are their rivals for nourishment cause them to starve. Bacteria may be killed before they die a natural death.

1. Death by Heat. Bacteria vary in their resistance to heat; some kinds are killed more easily than others. Spores are most difficult to kill. Food processors use heat in canning, pasteurizing, curing, and smoking of food. In *canning,* the food is subjected to a harsh heat treatment; the heating must be sufficient to kill the spores of the deadly food-poisoning organism, *Clostridium botulinum.* This organism grows best without air in a sealed can or jar. In *pasteurizing,* milder heating is applied. A good example of a pasteurized product is milk. Pasteurization procedures are carefully worked out to kill the majority of contaminating microorganisms and all the pathogens without destroying important quality characteristics of the food. However, pasteurized products do not necessarily possess sterility. Another example of a pasteurized product is egg; the contents of shell eggs must be pasteurized before they are frozen or dried.

Ham is heat treated during the process of *tenderization,* which destroys many, but not all, contaminants. This is an important fact to remember when storing tenderized hams (see Part Three, Section D).

Foods may or may not be rendered free from contaminants in the *cooking* process; the type of organism, the nature of the food, and the cooking procedure used, all have an effect. A detailed discussion of this subject will be found in Part Three, Sections F and G. Temperature control is an

*See the Readings at the end of Part Two.

extremely important tool in rendering and keeping food microbiologically safe.

Killing heat is also used—in the form of hot water—to *sanitize* eating and food-preparation utensils. The procedures used are discussed in Part Three, Section B.

Figure 4 illustrates the effect of temperature on bacteria.

2. Death by Chemical Agents. Chemicals can and sometimes are used as sanitizing agents in dishwashing and in treating work surfaces such as boards, tabletops, and counters.

MOLDS

Botanically, molds belong to a group called fungi. Fungi include many forms, from one-celled, microscopically small organisms to large structures called mushrooms. Unlike green plants, fungi lack chlorophyll.

Molds are a frequent cause of food spoilage and certain molds form toxins

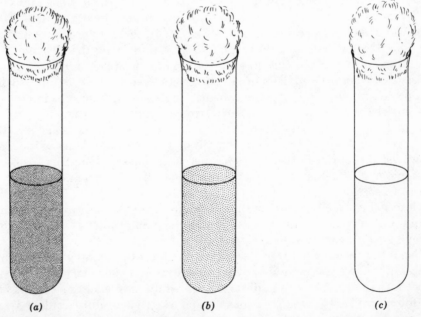

(a)	(b)	(c)

Figure 4. Effect of temperature on bacteria. *(a)* Growth is good or excellent between 45 and 140 F (7.2 and 60C); *(b)* Growth is very slow below 45 F (7.2 C). *(c)* Boiling kills vegetative cells, but spores and toxins may survive.

that have been shown to be injurious to test animals, and may have the potential of being toxic to human beings as well. Some molds when inhaled may cause respiratory ailments and may affect a person's eyes and ears. Outbreaks of illness in animals that had been fed moldy grain and the cancer-causing capacity of some of the fungal toxins have stimulated research not only in connection with moldy animal feed but with human foods also. Certain molds used by the dairy industry in processing Blue cheese, Roquefort and Gorgonzola are considered safe to eat.

Most molds are thread-like and the threads bear spores (Figure 5). Molds can be seen as a fuzzy, cottony growth of a variety of colors, on a great variety of foods. Molds can grow on practically all foods—the sweet, the sour, the bitter, the moist, and the dry. They are not choosy regarding temperatures at which they grow. Everyone is familiar with the appearance of mold on refrigerated food, around refrigerator doors, and on damp cellar walls; and of luxurious mold growth on shoes, books, and the like when the weather is hot and humid.

Mold spores are carried about in the air. One little piece of moldy bread

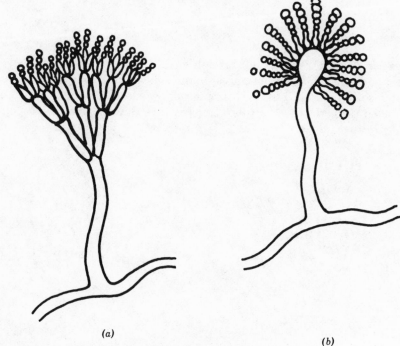

(a)

(b)

Figure 5. Molds. (a) Penicillium. (b) Aspergillus.

can contaminate other foods stored far away. Heating to the boiling point kills molds, including the spores. Mold spores serve the purpose of reproduction. In contrast to bacterial spores that are designed to withstand cold, drought, and other unfavorable conditions, mold spores are fairly readily killed by heat.

The toxicity of moldy feed in animals should serve as a warning to persons engaged in the foodservice business. Except for the molds used in the processing of a food (for example, in some cheese), there is no excuse for a food to be moldy. The presence of food spoiled by mold indicates poor management.

YEASTS

Yeasts possess microscopically small, thin-walled cells that multiply by sprouting (Figure 6a). They like sweet liquid foods and are well known for spoiling cider and fruit juices by fermenting them. Yeasts are purposely used in making bread, wine, and beer. From a public-health point of view, they are harmless. They can be readily killed by boiling.

VIRUSES

Viruses can also be transmitted through food. Viruses are smaller than the smallest bacteria. They cannot multiply in foods; in fact, they multiply only in living cells of their hosts. Examples of viruses known to have been transmitted through food are those causing infectious hepatitis and polio.

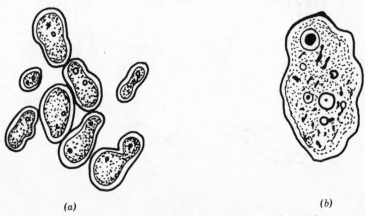

(a)   (b)

Figure 6. Yeasts and protozoan. (a) Yeasts. (b) A protozoan, member of the protozoa.

How does a virus such as the hepatitis virus find its way into food? First, from a person who has hepatitis or has recently recovered from this illness. Persons recovering from hepatitis are apt to be carriers for a while (a carrier is a person who appears to be healthy, but who still harbors the germs of the disease with which he or she was ill). When a hepatitis carrier visits the toilet, does not wash the hands thoroughly and then handles food, the person contamintes the food with the virus clinging to his or her hands.

Another source of viruses is sewage-polluted waters (such as streams, lakes, and oceans) and the fish and shellfish harvested from them. A third source may be raw sewage that is allowed to drop onto food or kitchen tables from leaky overhead pipes, or through otherwise faulty plumbing.

PROTOZOA

Protozoa are very small animal-like microorganisms (Figure 6*b*) that are widely distributed in lakes, ponds, streams, oceans, and in moist soil rich in organic matter. Protozoa may occur in the intenstinal tract of humans. Several of these organisms cause intestinal disorders, the so-called amebic dysenteries. The amebae may be transferred to food with fecal matter

Figure 7. Trichinae.

clinging to the hands of food handlers who have dysentery or have had the disease and have become carriers of the organism. This shows the necessity of thorough hand washing after a visit to the toilet.

Vegetables and fruits that have had contact with contaminated soil may become contaminated. Fortunately, in the U.S.A. dysentery-producing amebae are not common contaminants of soils on which truck crops are raised.

TRICHINAE

Trichinae are microscopically small, wormlike parasites that cause a disease in hogs, rabbits, and bears. The trichinae settle in the muscle meat of the hosts (Figure 7). When we eat undercooked, infested meat, we may also become ill with trichinosis.

Hogs contract the disease by eating uncooked, infested garbage. Fortunately, trichinosis of hogs becomes less prevalent every year. However, one must not take chances. It is important to use proper cooking methods for all pork, rabbit, and bear meat, and to heat these meats to the grey state. Thorough heating kills the trichinae and renders the meat safe for human consumption.

Cutting and grinding of infested raw meat may contaminate cutting boards and grinders; therefore, these surfaces should be thoroughly washed after each use.

SECTION C

Definition of Terms

Acid is the reaction of a substance having a pH below 7.

Aerobic is a microorganism needing free oxygen for its existence and reproduction.

Alkaline is the reaction of a substance having a pH above 7.

Anaerobic is a microorganism preferring the absence of free oxygen for its life and reproduction.

Antiseptic is the condition of being free of, or cleaned of microorganisms.

Bacilli are rod-shaped bacteria capable of forming spores.

Bacteria are plantlike single cells that are only visible with the aid of a microscope; they multiply by splitting in two.

Bactericidal is a term that describes a substance capable of killing bacteria.

Bactericide is an agent capable of killing bacteria.

Bacteriostatic is a term that describes a substance that inhibits the multiplication of bacteria.

Botulism is a frequently fatal type of food poisoning caused by a toxin produced by a spore-forming bacterium under exclusion of air; the bacterium is *Clostridium botulinum.*

Budding is a form of reproduction by which the mother cell produces a new cell by forming an outgrowth resembling a bud. Yeasts multiply by budding.

Carrier is a person who harbors organisms causing a communicable disease without showing outward signs of illness.

Celsius is the name of an astronomer who devised the centigrade scale of temperature. In this scale 0° represents the ice point and 100° the steam point. Centigrades are converted to degrees Fahrenheit by the formula: $(°C \times 9/5) + 32 = °F$. (See also Fahrenheit).

Chilling is cooling above the freezing point. Food is usually chilled by refrigeration.

Clostridium botulinum is an anaerobic spore-forming rod-shaped bacterium; it is the cause of botulism.

Clostridum perfringens is an anaerobic spore-forming, rod-shaped bacterium; it is the cause of perfringens food poisoning.

Coccus (cocci, plural) is a round bacterium.

Coliform bacteria are undesirable in foods. Their presence in some foods, especially seafood, milk, and water, indicates the possibility of contamination with excrement (manure, sewage). One of the species of the group, *Escherichia coli,* is a common inhabitant of the intestinal tract of humans and animals.

Communicable diseases are contagious or "catching" diseases. The causative agents are readily transferred from person to person.

Contagious means capable of being transmitted by contact with a person or object.

Contaminated food is food that contains undesirable microorganisms; contamination is the act of contaminating.

Cross contamination is the contamination of clean, wholesome food with illness-producing bacteria adhering to raw food, soiled hands, soiled work surfaces, soiled equipment, soiled utensils, soiled clothing, and the like.

Crustacea is the class of animals comprising the crustaceans; these include lobsters, shrimps, and crabs.

Danger zone of bacterial growth is the temperature range from 45 to 140 F (7.2 to 60 C). Within this range, growth of food-poisoning bacteria is possible. Food temperatures should not be within this range for longer than 4 hours, and should not be within the 60 to 120 F (15.6 to 49 C) range for longer than 2 hours.

Detergent is a water-soluble cleaning agent, a synthetic organic compound having wetting and emulsifying properties. It is soaplike in action, but chemically not a soap.

Diplococci are cocci hanging together in pairs.

Disinfectant is a chemical agent used to inhibit the growth of, or to destroy, harmful microorganisms.

Entameba histolytica is a member of the protozoa, causing amebic dysentery.

Enzymes are complex organic substances originating from living cells.

Epidemic is a disease that affects a large number of people at the same time; the disease is largely spread from person to person.

Facultative are microorganisms that grow well either aerobically (free oxygen available) or anaerobically (free oxygen not available).

Fahrenheit is the name of a physicist who devised a temperature scale in which 32° represents the ice point and 212° the steam point. Degrees Fahrenheit (F) can be converted to degrees Celsius (C) using the formula: $(°F - 32) \times 5/9 = °C$.

Fission is a method by which vegetative bacterial cells multiply by splitting, with each portion developing into a new cell.

Flagella are whiplike appendages on the vegetative cells of some forms of bacteria. They facilitate locomotion for short distances only. The flagella originate from one or both ends of the cell, or are distributed all over the cell; their position depends on the kind of bacterium.

Food handler is a person who handles food in the foodservice establishment regardless of whether the person actually prepares or serves food.

Food infection is an illness caused by eating food that contains bacteria capable of causing a gastrointestinal infection in the victim.

Food intoxication is a foodborne illness caused by a toxin, or toxins. The toxin, or toxins, may be of bacterial origin.

Food poisoning–see Food intoxication. The term is frequently widened to include food infections.

Food spoilage is the deterioration of food, making it unfit for human consumption.

Foodborne illnesses are caused by the ingestion of food contaminated with disease-producing agents such as bacteria and parasites; or foods that by their nature are poisonous to humans, such as poisonous fish, mushrooms, or other poisonous plants.

Galvanized means coated with zinc.

Gastroenteritis is an inflammation of the stomach and the intestinal tract.

Gastrointestinal pertains to the stomach and the intestinal tract.

Generation time is the time that elapses between one division of a bacterial cell into two daughter cells and the next division of each of these daughter cells.

Germicide is an agent used for killing microorganisms.

Germination is the process of forming a sprout. Bacterial spores (thick walled) will germinate, forming a vegetative cell which by subsequent division, gives rise to a new crop of thin-walled cells.

Growth curve is a curve that depicts the multiplication of bacteria over a period of time.

Growth of bacteria denotes the multiplication of bacteria rather than their increase in size.

Hepatitis is an inflammation of the liver caused by a virus that may be transmitted by food and water.

Hygiene, personal, refers to the cleanliness of one's body.

Immunization is the process of rendering an individual immune, or resistant, to a disease.

Incubation time refers to the period between infection and the appearance of signs of the disease.

Infections, foodborne, are illnesses caused by swallowing certain pathogenic bacteria present in the food. Food infections may be caused by relatively few cells of highly infectious pathogens that are transmitted by food; or by pathogens that must multiply in the food to high numbers before they make the victim ill.

Infectious refers to the ability of a microorganism to produce an infection in a susceptible host.

Inhibitors of microorganisms are substances that interfere with their growth.

Intoxications, foodborne, are illnesses by a toxin, or poison in the food consumed; the toxin may be of bacterial origin.

Iodophors are iodine compounds used as bactericidal agents in sanitizing food-contact surfaces.

Lag phase of bacterial growth is the inactive period preceding bacterial mutliplication; the cells are there, but they are not yet dividing.

Log phase of bacterial growth is the period of constant and rapid multiplication; the generation times are shortest during this period.

Mesophilic bacteria multiply most profusely at moderate temperatures.

Micron, one-millionth of a meter, is a unit of measurement useful in expressing the size of very small objects, such as bacteria.

Microorganisms are microscopic plants or animals.

Milliliter is a very small measure; it equals 1/1000 of a liter (or approximately 1/1000 of a quart). The milliliter is used to express bacterial counts in fluids such as water and milk.

Multiplication of bacteria, or bacterial growth, takes place by the splitting of a bacterial cell into two, which in turn split into two, and so on.

Neutral, 7 on the pH scale, expresses the borderline between acidic and alkaline.

Optimum means best.

Outbreak, foodborne, is the occurrence of foodborne illness in two or more people who have eaten a food shown by proper analysis to be the cause of the illness.

Parasites are organisms that survive in living hosts, from which they draw their nutrients. Animals and plants can be parasitic. The latter are not important in connection with foodborne illnesses.

Pasteurization is a process of killing microorganisms in a food (for example, milk) by heating it at well-controlled temperatures (which are considerably below boiling) for specified lengths of time.

Pathogen is an organism causing disease.

Pathogenic means causing disease.

Perishable is a food that becomes readily unfit for human consumption. The change may be caused by the action of enzymes naturally present in the food, or by enzymes released by microorganisms present in the food.

pH expresses the acidity or alkalinity of a substance. A pH of 7 indicates that the substance is neither acid nor alkaline. As the pH values increase to 14, the alkalinity increases; as the pH values decrease to 0, the acidity increases.

Pollute means to render foul, unclean, dirty, or contaminated with fecal matter.

Potable means safe for drinking.

Potentially hazardous foods are foods that consist entirely, or in part, of milk, eggs, meat, poultry, fish, or shellfish; or that are for other reasons capable of supporting the growth of microorganisms causing foodborne illness. Many much-handled, lightly heated, or unheated foods are potentially hazardous. (*See* pg. 13).

Protozoa are small animal-like microorganisms; some are transmitted through food.

Psychrophilic bacteria are cold-loving bacteria.

Ptomaine is a poorly defined group of bacterial toxins that in the past was often used to describe a cause of food poisoning that was most probably a staphylococcal intoxication.

Putrefaction is the process of becoming putrid or rotten.

Putrid means rotten or foul-smelling.

Quatenary ammonium is a substance present in many sanitizing compounds.

Recontaminating means contaminating anew food or surfaces previously rendered free from bacteria.

Respiratory tract is the part of the anatomy that serves the function of breathing or respiration.

Reservoir is a term to denote a passive source or carrier of a pathogen.

Route of disease transmission—see Vehicle of transmission.

Safe temperatures for food are temperatures at which microbial growth is very slow or nil: 45 F (7.2 C) and below; and 140 F (60 C) and above.

Salmonellae are rod-shaped, facultative nonsporeformers that are capable of causing salmonellosis in human beings and animals.

Salmonellosis is an infection of the gastrointestinal tracts caused by the ingestion, in large numbers, of certain species of *Salmonella*.

Sanitation of food pertains to the wholesomeness and cleanliness of food.

Sanitizing means reducing the number of microorganisms on articles or in substances to levels judged safe by the health authorities. In food service, sanitizing refers to treatments to eliminate pathogenic bacteria from equipment and utensils that are used in the preparation and service of food.

Solvent is a substance that dissolves another to form a solution.

Species refers to one kind of microorganism; a species is a subdivision of a genus. Therefore, a genus is a group of closely related species.

Spoilage of food refers to the deterioration of food, which makes it unfit for human consumption. However, people vary in what they regard as deteriorated or unfit to eat.

Spores are thick-walled bodies formed by some microorganisms, such as molds and certain bacteria. Spores are resistant to heat, drought, and other unfavorable conditions. The soil abounds with spores.

Staph is an abbreviation of staphylococci.

Staphylococci are cocci bunched in grapelike clusters.

Stationary phase of bacterial growth is the phase following the log phase (and the phase of very rapid growth) during which the number of cells remains fairly constant.

Sterile means free of living organisms.

Streptococci are cocci arranged in chains.

Thermoduric bacteria are bacteria capable of surviving exposure to heating processes such as pasteurization and canning.

Thermophilic bacteria are heat-loving bacteria.

Time-temperature control of bacteria in food refers to regulating the length of time for which a food is allowed to remain at certain unsafe temperatures. At unsafe temperatures, bacteria may multiply rather rapidly. At safe temperatures, growth is halted.

Toxic foods are foods that contain toxins; these toxins may or may not stem from microbial growth.

Toxicogenic refers to the ability to produce a toxin.

Toxins are poisonous compounds.

Trichinae are small wormlike parasites embedded in pork, rabbit, and bear meat and originating from animals suffering from trichinosis.

Trichinosis in humans is a type of foodborne illness following the eating of trichinae-infested undercooked pork, rabbit, or bear meat.

Vegetative bacterial cells are thin-walled cells that multiply by splitting.

Vehicle of transmission refers to the route by which a pathogen is passed on from its original source to a victim. Vehicles of transmission can be soiled hands, sewage, water, air, food, rats, mice, flies, cockroaches, and soiled food-preparation equipment and utensils.

Vibrio is a genus characterized by comma- or S-shaped bacteria; certain species are pathogenic to humans.

Vibrio parahaemolyticus is a member of the genus *Vibrio*. It is a cause of foodborne infection in human beings. The organism is associated with sea water and seafood harvested from contaminated waters.

Viruses are very small pathogens that multiply in the living cells of their hosts; they do not multiply in cooked food.

Water activity expresses available moisture in contrast to total moisture. The symbol for water activity is a_w.

Wholesome food is sound food, fit for human consumption.

Yeasts are microorganisms that multiply by budding and may cause spoilage of fruit juices by fermenting sugar to alcohol and carbon dioxide. Yeasts are used in preparing yeast-leavened doughs, wine, and beer.

Yersinia enterocolitica is a pathogen capable of causing a foodborne infection, yersiniosis, in humans. It is able to grow at low (refrigerator) temperatures.

Important Facts in Brief for Part One

- Bacteria are very small and are everywhere.
- Certain kinds of bacteria are harmful to humans; some cause food spoilage, others cause disease.
- Bacteria causing communicable (or catching) diseases can be transmitted through food.
- Certain so-called food-poisoning bacteria may cause illness if allowed to multiply in food to high numbers, after which the food is served and eaten.
- Some forms of food-poisoning bacteria release toxins into food.
- Bacterial spores are very hardy and survive harsh heat treatments and other adverse conditions.
- Vegetative bacterial cells reproduce by splitting in two; this process is very rapid when conditions are favorable.
- Bacteria need food, moisture, and warmth for growth. With respect to the safety of foods, the temperature range in which growth is possible is called the DANGER ZONE. This zone ranges from 45 to 140 F (7.2 to 60.0 C). Growth is excellent between 60 and 120 F (15.6). and 49 C).
- Bacteria can be killed by heat; to kill spores requires more heat than to kill the delicate, vegetative cells.
- Heat is used to kill pathogenic bacteria in canning and pasteurizing.
- Cooking temperatures do not always suffice to kill pathogenic bacteria.
- Molds have not been indicted in food-poisoning outbreaks. However, certain molds form toxins that have been shown to cause cancer in test animals.
- Viruses can be transmitted by food and water. Hepatitis virus has caused well-documented food-poisoning outbreaks.
- Protozoa are animal-like microorganisms. Certain members may cause food- or waterborne dysenteries in human beings.
- Trichinae, a not uncommon parasite of hogs, can be transmitted through undercooked pork to people, causing trichinosis.

Study Questions for Part One

1. In your own words, explain what is meant by sanitation as it pertains to food service.
2. Define the term "microorganisms"; give examples of microorganisms that are important in connection with food sanitation.
3. Describe the growth curve of bacterial multiplication.
4. Define pasteurization.
5. Can viruses multiply in cooked foods? Explain.
6. What are protozoa? Can protozoa be transmitted through food?
7. Define trichinosis.
8. Explain psychrophilic, thermophilic, and thermoduric.
9. Can bacteria be killed by heat? Explain.
10. Are bacteria killed when food is frozen? Explain.
11. What are important differences between vegetative bacterial cells and bacterial spores? Explain.
12. Define the terms "aerobic," "anaerobic," and "facultative" bacteria.
13. In what way are bacterial spores different from vegetative cells?
14. In what way does the pH of a food affect bacterial growth?

Readings for Part One

Burdon, K. L. and R. P. Williams. *Microbiology.* 6th ed. Macmillan, New York. 1968.

Frazier, W. C. and D. C. Westhoff. *Food Microbiology.* 3rd ed. McGraw-Hill, New York, 1978.

Jay, J. M. *Modern Food Microbiology.* Van Nostrand, New York. 1970.

Longrée, Karla. *Quantity Food Sanitation.* 3rd ed. Wiley-Interscience. New York. 1980.

Pelczar, M. J., R. D. Reid, and E. C. S. Chan. *Microbiology.* 4th ed. McGraw-Hill. New York. 1977.

Stevenson, G. *The Biology of Fungi, Bacteria and Viruses.* 2nd ed. Edward Arnold Publishers, Ltd., London. 1970.

PART TWO

Food Spoilage and Foodborne Illnesses

SECTION A

Food Spoilage, Defined

What is food spoilage? Spoilage refers to food decomposition as well as to other forms of damage. Food spoilage can cause considerable financial loss to the foodservice business, but it does not necessarily represent a health hazard. Much spoiled food is not eaten and is simply thrown out. Spoilage may be recognized by the food's appearance, smell, taste, consistency, or a combination of these. However, foods that harbor food-poisoning bacteria or other pathogens usually appear to be sound and good to eat, although they are actually dangerous to health.

Spoilage denotes that a food is "unfit to eat." Unfitness to eat is a loose term, since the consumer's ethnic background and personal preference affect his or her judgment. Yet, certain standards do exist for foodservice managers entrusted with the purchase of food items.

Foods fit to eat:

a. Have the desired stage of development or maturity.
b. Have been free from pollution throughout their production and subsequent handling.
c. Are free from signs of physical and chemical action that decreases the quality of a food. This action is caused by enzymes (which cause overmaturity and finally rot); activity of microorganisms, rodents, insects, and parasites; and damages from mishandling, such as exposure to heat, cold, drying, jostling, too tight or too loose packaging, and other forms of mishandling.
d. Are free from microorganisms and parasites capable of causing foodborne illnesses.

About one-fourth of the world's food supply is lost through microbial activity alone. Added to this must be the losses caused by overmaturity, insect infestation, and rodent activity. Rats in particular may cause tremendous losses because of their voracious appetite and ability to bring forth new generations of young in rapid succession. Rodents also are notorious carriers of pathogens.

In a foodservice establishment, the loss of food through spoilage can cause an enormous economic burden—in most instances, an unnecessary one. Much spoilage is due to poor judgement of the food buyer, such as

- Failure to keep abreast of market forecasts
- Overbuying
- Failure to write accurate specifications
- Insufficient inspection of incoming commodities
- Lack of promptness in storing delivered items
- Inadequate storage facilities
- Lack of frequent inventory
- Failure to separately store dairy products, meats, vegetables, and prepared foods
- Failure to maintain refrigerators and freezers in good operating condition
- Failure to maintain storage areas in a sanitary condition at all times
- Insufficient training and supervision of personnel.

It is all a question of knowledge and good management.

Most foods will eventually spoil, and the so-called perishables spoil most readily. For the sake of convenience, we may classify foods on the basis of their perishability or stability as follows:

1. Nonperishable or Stable Foods   These foods do not spoil unless they are handled without care. Examples are sugar, macaroni, dry beans, flour, spaghetti, noodles, and other staples.

Even stable foods may spoil when they are improperly stored and are accessible to insects and rodents. Storage of dry food in tight containers is a must. The storage area must be kept neat and clean. The storage area for nonperishables is known as "dry storage." The humidity in this area should be low; a damp atmosphere invites the growth of microorganisms, especially molds. Damp conditions also create musty odors that will be absorbed by the food stored. Good ventilation is essential and lighting should be sufficient to allow for inspection of the food and inventory. Sunlight, however, should be excluded, and windows protected against entrance of insects and rodents. Temperatures from 40 F (4.4 C) to a maximum of 70 F (21.1 C) are adequate.

The requirements for storage conditions of commodities are more fully discussed in Part Three under "General Rules for Checking and Storage of Deliveries."

2. Somewhat (semi) Perishable Foods   These items will remain free from spoilage for a long time if properly protected from adverse conditions. Examples are nuts, potatoes (white and sweet), onions, and waxed vege-

tables (for example, rutabagas); frozen foods, provided they are kept solidly frozen at 0 F (−18 C); and canned foods, if stored in cool and dry storage rooms.

Although they are semiperishable, these items need most careful attention in selection and storage. Nuts should come from the current year's crop. They should be bought in limited quantity since they become easily infested with insects; also, the flesh may become rancid rather fast. Cool storage for a short time is advised.

Potatoes and root vegetables should be carefully inspected for firmness and freedom from injuries caused by rough handling or microbial activity. After thorough check for quality, which should be specified when purchasing, store root vegetables in an area designated for the sole storage of vegetables. Do not overbuy. Onions and sweet potatoes spoil easily, should not be stored for more than 3 to 4 weeks and checked frequently for signs of decay. White potatoes and rutabagas should keep up to 3 months under cool (60 F; 15.6 C) conditions. Spoilage can be detected by sight and smell. One rotten potato or onion may form a nucleus for an entire colony of rotten items creating a nuisance in the storage room by an unmistakable stench. Remove suspect items at the first sign of decay.

Frozen foods must be expected to contain bacteria, since freezing does not kill all bacteria by any means. When frozen foods are defrosted, they are again highly perishable. Bacteria injurious to health may begin to multiply when the **DANGER ZONE**, which begins at 45 F (7.2 C), has been entered.

Canned foods have excellent keeping quality because the contents of the sealed cans are practically free from bacteria since the dangerous bacteria are removed in the canning process. Some heat-loving and hard-to-kill thermoduric spores of certain bacteria may remain, however. As we learned earlier, heat-loving bacteria are able to grow at warm temperatures. We also learned that certain bacteria grow without air. Therefore, if spores of heat-loving, air-hating bacteria are present in canned food and the can is stored at very warm temperatures, such as 113 F (45 C) and higher, these spores may resume growth and the contents of the can may spoil. Other types of spoilage of canned goods are also possible; for example, when a can has a leak, bacteria may penetrate into the can and spoil the food. Contents of purposely opened cans may also spoil if stored improperly or for too long.

Signs of spoilage of canned foods are:

(a) Cans that are leaky, corroded, or rusty.
(b) Cans that are bulgy.
(c) Contents that spurt out when opening can.
(d) Contents that appear bubbly, slimy, decomposed, or moldy.
(e) Contents that smell putrid or sulphurous.

Note: Contents may be poisonous. Do not taste from a can of doubtful wholesomeness. Discard it.

3. **Perishable Foods** This group is a large one and contains most of our daily foods, such as milk and other dairy foods, meat, poultry, eggs, fish, shellfish, fruits, and vegetables. All prepared menu items are perishable; so are canned foods once the can has been opened and frozen items once they have thawed.

What Are the Signs of Spoilage?

The signs of spoilage of perishables vary with the different types of food, the cause of spoilage, and the environment. Spoilage can be caused by overmaturity, the activity of microorganisms, frost, heat-drying, excessive humidity, or combinations of various causes. The visible signs can be softening, hardening, leaking, discoloration, shriveling, moldiness, or bubbling. There can be off-odors such as moldy, yeasty, putrid, musty, and sulphuric. And there can be off-flavors such as rancid, sour, alcoholic, bitter, yeasty, and putrid.

Spoilage of fruits and vegetables can occur very fast, within several days, even hours. Therefore, fruits are usually purchased before the stage of full maturity is reached; they should be free from spots, cracks, and other blemishes. Fruits are kept either at room temperature for fast ripening; or refrigerated at 40 to 45 F (4.4 to 7.2 C) if ripening is to be delayed, but not longer than up to 5 days. Vegetables, except root vegetables, should be used promptly; when stored, they should not be washed before storing. A storage period of 5 days is maximum. Vegetables should also be devoid of spots and other signs of poor quality.

Although the microorganisms causing spoilage of fruits and vegetables are not pathogenic to humans, they are sometimes the cause of considerable economic losses. Also, microbial spoilage causes unsanitary conditions in the storage rooms that result from the creation of decayed matter and sometimes noxious odors. Spoilage, in whatever form it occurs, is a sign of sloppy management.

Spoilage of proteinaceous foods is not only an esthetic nuisance and economic loss, but it can also be a serious threat to public health. Meat, poultry, and seafood are apt to carry food-poisoning organisms on them, and conditions favorable to spoilage organisms are usually conducive to multiplication of pathogens as well. Discolored or moldy meat and meat products, slimy or flabby poultry, slimy or flabby fish with sunken eyes and grey-green gills—all are suspect and should be discarded. Signs typical of potentially hazardous items are of particular interest from a public health point of view. Therefore, they have been presented in a later chapter of the book (Part Three, Section D) where purchase and proper storage of potentially hazardous items are discussed. Also, books dealing with the topic of food purchasing in general should be consulted; some are listed under Readings for Part Two.

Cooked foods may show the various signs of spoilage listed above; or they may not show any signs and yet may have become dangerous, sometimes within hours, due to the activity of microorganisms capable of causing food intoxication (poisoning) or gastrointestinal infections. This treacherous fact must be remembered by all who prepare and serve food. Since there frequently is no noticeable evidence of spoilage, one cannot rely on signs to judge cooked items for wholesomeness. One must plan, prepare, hold, and serve food in a way that assures its wholesomeness. One must exercise certain definite precautions in the care of leftovers (See Part Three, Section F and G).

Foodborne Illnesses,
Defined

In the following presentation foodborne illnesses are divided into three groups. It is important that the basis for this grouping is understood before the reader proceeds.

1. ILLNESSES CAUSED THROUGH TRANSMISSION OF DISEASE GERMS

The food may transmit pathogenic germs from one person to another, much as germs are passed from person to person through other soiled objects such as money, doorknobs, railings, common drinking cups, and the like. Food serves as a mere vehicle of disease transmission. Transmission of animal pathogens to humans by way of food is also possible.

2. FOOD INTOXICATIONS AND FOOD INFECTIONS CAUSED BY BACTERIA

The term food poisoning refers to a violent illness of the stomach and intestinal tract (known as gastroenteritis) following the consumption of the offending food.

The offending food contains, in general, high numbers of bacteria that are capable of producing the gastroenteritis suffered by the victim. Some of these pathogens are able to release toxins into the food; these toxins are the direct cause of the illness, an intoxication. Other pathogens do not act until swallowed, whereupon they cause an infection of the gastrointestinal tract.

3. FOOD POISONING CAUSED BY AGENTS OTHER THAN MICROORGANISMS

The offending food contains poisonous chemicals, or the food is a poisonous plant or animal. The food could also contain injurious particles, such as chips of glass and china, pieces of metal, wood, or sharp bone—all of which may injure the customer's mouth when he or she accidentally bites down onto such materials and may injure other parts of the body after the person swallows them.

SECTION B

Transmission of Pathogens Through Food

I. TRANSMISSION OF PATHOGENS OF HUMAN ORIGIN

What are these diseases?

Most of the diseases transmitted by food are known to be those of the human respiratory and the intestinal tract. Important transmissible diseases are the common cold, septic sore throat, sinus infection scarlet fever, tonsil and adenoid infections, pneumonia, diphtheria, tuberculosis, influenza, typhoid and paratyphoid fevers, the dysenteries, cholera, and infectious hepatitis. The organisms causing the diseases are bacteria, protozoa, or viruses. See Tables I and II in Appendix A.

How are they transmitted from man to food? (Fig. 8 and 9).

Ill Food Handler

Infected sores, often starting as cuts or burns, are a rich source of staphylococci. A diseased respiratory tract is another important source of pathogens. Respiratory illnesses are commonly transmitted through discharges from mouth and nose while coughing and sneezing. Hands and handkerchiefs soiled with discharges from the mouth and nose are also sources; so are tasting spoons used more than once without cleaning. Displayed food can be contaminated by the coughs and sneezes of customers as well as those of personnel. Pathogens causing diseases of the intestinal tract can be transmitted by food when a food handler, ill or acting as a carrier, does not wash, or not thoroughly wash, hands after visiting the toilet. Examples of intestinal diseases are typhoid and paratyphoid fevers, amebic and bacillary dysenteries, and infectious hepatitis.

Principles of Control. Persons acutely ill with respiratory, intestinal, or other contagious disorders, must not handle food. Carriers of pathogens must not handle food. An employee hired for work where he or she touches

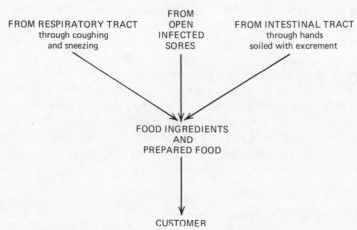

Figure 8. Direct transmission of microorganisms from food handler to food and customer.

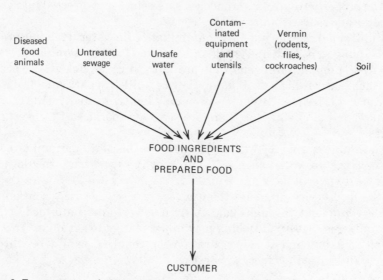

Figure 9. Transmission of microorganisms from various sources to food and customer.

food and food-contact surfaces such as equipment and utensils should come with a clean bill of health. If the employee has been ill with a communicable disease while employed, he or she should not be readmitted to work until a physician has attested to the individual's freedom from pathogens that caused the disease. More will be said about the subject in a later chapter (Part Three, Section E).

What other vehicles of transmission are there?

Besides the immediate transmission of pathogens from the food handler or customer to prepared food and from there to the next person or persons, additional sources and routes of transmission are possible. These are illustrated in Figure 9. The various sources include diseased food animals, untreated sewage, unsafe water, contaminated equipment and utensils, and vermin such as rodents, flies, and cockroaches.

Sewage and Sewage-Polluted Water, Soil, and Food

Sewage consists of people's wastes. Therefore, raw, untreated sewage must be expected to contain pathogens that have been eliminated from the human body. Examples are the germs causing typhoid and paratyphoid fevers, dysentery, and infectious hepatitis. If raw sewage is allowed to drain into water bodies such as wells, rivers, and lakes used as sources of water for drinking and food preparation, a serious hazard is created. Privies and septic tanks should be sufficiently removed from wells, and raw sewage should be properly treated and rendered safe before it is directed into rivers, lakes, or other sources of drinking water.

Raw untreated sewage not only contaminates the water itself but also the animals living in it. Fish and shellfish harvested from contaminated waters may harbor human pathogens that are able to cause disease in persons eating such hazardous foods. Contaminated shellfish is particularly danger-ous because it is frequently eaten raw or only lightly heated during cooking. The harvesting of shellfish is therefore strictly supervised by government agencies.

If raw sewage is used to fertilize fields, the soil becomes contaminated and so do the crops grown on such soil. This is why raw sewage should not be used in raising vegetables and fruit.

On the foodservice premises themselves, raw sewage may contaminate food and equipment through leakage from sewer pipes onto food or food-contact surfaces. Faulty plumbing may allow sewage to get mixed in water used for drinking and food preparation, or could contaminate, through backflow, sinks used for food preparation.

Principles of Control.   Purchase food (in particular, shellfish) from sources approved by the health authorities (see Part Three, Section D).

Rodents and Insects

Rodents, especially rats, are notorious for harboring pathogens and will carry these into the foodservice establishment if permitted entrance to it. The presence of rodent excreta is a sure sign of the presence of these animals. Rats and mice may carry pathogens on their feet, in their fur, and in their intestinal tract. Rats love to roam and feed in garbage dumps and sewers. Rodents also harbor salmonellae, which are capable of causing food poisoning.

Principles of Rodent Control.   Rodents must not be tolerated in a foodservice establishment. They must be eradicated professionally and their reentry must be prevented (see Part Three, Section A).

The housefly is the type of fly commonly associated with living quarters and eating places, but also with toilets, barns, manure, garbage, and other filth. This crisscrossing from filth to food makes the fly a menace in the foodservice establishment.

As the fly feeds on a food, it periodically regurgitates liquid from its crop to dissolve the food for easier uptake. This regurgitated liquid may contain disease-producing germs. The fly also carries germs on its feet.

Cockroaches are common pests in kitchens and bakeshops. Like the flies, they contaminate food with their mouth parts, feet, and droppings. They, too, regurgitate filth onto clean food while they feed on it.

Principles of Control.   Flies and cockroaches must be eradicated and kept out of the areas where food is prepared and served. Also, they must not be permitted access to toilets and garbage (see Part Three, Section A).

Contaminated Equipment and Utensils

Disease-producing organisms can be transmitted through equipment and utensils if these have been touched by ill, or carrier-type, food handlers and customers; or visited by flies and cockroaches; or exposed to dust or to dripping sewage; or to sewage that has backed up into sinks or other drainage pipes due to faulty plumbing; or through contact with contaminated water and food.

Principles of Control. Equipment and utensils must be protected from becoming contaminated through contact with the above sources of contamination. They must be maintained in a sanitary condition at all times (see Part Three, Section B).

II. TRANSMISSION OF ANIMAL PATHOGENS

What are these diseases?

Only diseases of present importance will be discussed. Great strides have been made to eradicate some diseases that in the past greatly threatened animals as well as humans, such as brucellosis. This is a disease of cattle that can be transmitted to humans through unpasteurized or improperly pasteurized milk.

Trichinosis is another disease that can be transmitted from animals. It is caused by a very small roundworm, *Trichinella spiralis,* which is a parasite of hogs, and may be transmitted to humans if meat containing live larvae of the parasite is consumed. This may happen when pork is undercooked. When infested meat containing live larvae is eaten, the larvae develop into mature worms in the victim. Initial symptoms may appear within 2 days, but they also may take much longer. These symptoms are much like those of other foodborne gastrointestinal disorders and include nausea, vomiting, and intestinal distress.

To assure that pork is rendered safe it should routinely be cooked to a gray color, that is, to an endpoint of 165 to 170 F (74 to 77 C).

The larvae can also be destroyed by freezing, if pork is freezer-stored at 5 F (−15 C) for 30 days; or − 10 F (−23 C) for 20 days; or − 30 F (−34 C) for 12 days. This same organism also causes trichinosis in bears and rabbits. Thorough cooking destroys the trichinae. For quick reference see Table III in the Appendix.

Tapeworms of animal origin (cattle, hogs) may be transmitted to humans if infested undercooked beef and pork are consumed. Infested meat is referred to as looking "measly." Carcasses are inspected for this condition before they go on the market and incidences of illness resulting from eating tapeworm-infested meat consequently are rare. However, if they do occur, such attacks can be serious. Therefore, eating raw beef (steak tartare) poses a potential danger to health, however remote the actual danger may be thanks to thorough inspection. Cooking heat destroys the parasite.

A disease of wild rabbits, tularemia, is of greater importance to the homemaker than to managers of quantity foodservice operations. In the U.S., wild rabbit is not apt to be served in commercial food service.

In summary, the principles of control of disease transmission by way of food are:

1. Persons ill with, or carriers of, contagious diseases must not handle food.
2. All food handlers must observe the rules of personal hygiene. They must thoroughly wash their hands after a sneeze or cough and after visiting the toilet.
3. The food and water supply should come from uncontaminated sources.
4. Rodents and insects must be strictly controlled in the foodservice establishment.
5. Equipment and utensils used in food preparation and service must be maintained in a sanitary condition at all times.

The contamination of food with microorganisms from the human and animal source to menu items during their preparation, holding, and service will be discussed in the next chapter.

For brief reference, the characteristics of pathogens of human and animal origin that may be transmitted to people by way of food are presented in Appendix A, Tables I, II, and III.

SECTION C

Food Poisonings and Food Infections Caused by Bacteria

Food poisonings and bacterial infections are usually called simply "food poisoning." These illnesses are characterized by a violent form of upset of the stomach and intestines (gastroenteritis), which follows the consumption of the food in which the offending bacteria were given a chance to multiply. *Bacterial food poisoning is the result of the mishandling of food.*

The time between the consumption of the food and the appearance of the symptoms (called the incubation time) may vary. The violence of the symptoms, such as nausea, vomiting, and cramps of the stomach and intestines (gastrointestinal tract), may vary also. Variation in incubation time and intensity of discomfort depend on:

1. The type of organism causing the illness; some cause a more severe illness than others.
2. The susceptibility of the person; the very young, the aged, the sick, and the feeble are more susceptible.
3. The number of bacteria or the amount of toxin consumed; the greater the number of bacteria or the greater the amount of toxin swallowed, the quicker and more severe the attack.

What is the difference between bacterial food poisoning and bacterial food infection?

True bacterial food poisoning is caused by a poison, or toxin, which the offending bacteria release into the food, thus rendering it poisonous. True food poisoning is also called a food intoxication. Bacterial food infection, however, is not caused by a toxin released into the food. Rather, the bacteria set up an infection in the gastrointestinal tract of the victim after having been swallowed by the victim. *In either case, a great number of bacteria must be present in the food before it will make a person sick.*

The presence of large numbers of bacteria in food indicates that something has gone wrong in the preparation of the offending food—namely, that is has been mishandled. "Mishandling" is a wide category and can refer to a variety of unsafe treatment administered to food, including poor time-temperature control.

The main purpose of this book is to awaken an understanding of the dangers associated with poor sanitary techniques used in food handling and to show when and how to use appropriate techniques in the production and service of safe food prepared in quantity.

We will now discuss some important types of foodborne intoxications and infections.

BOTULISM—A FOOD INTOXICATION

Botulism is caused by one of the most powerful poisons, the toxin of a bacterium, *Clostridium botulinum*. The toxin attacks the nervous system and the victim usually dies because he or she cannot breathe properly. Botulism is now rare. For quick reference see Table IV in the Appendix.

The bacterium causing botulism is a rod capable of forming spores. It is gas-forming, anaerobic, and grows well over a wide range of acidity with a pH range of about 5–8. Seven types of botulism can be distinguished on the basis of the toxin produced:

Type A is most common in the soils of the Western part of the USA. It is a very toxic strain.

Type B is very widespread in soils all over the world. It is less toxic than type A.

Types C and D cause botulism in certain animals, but not in human beings.

Type E usually occurs in the bottom of streams, lakes, and oceans and is therefore associated with seafood.

Type F is similar to types A and B. It may cause botulism in people, but it is rare.

Type G has been recovered from soil in Argentina. It has not been shown to cause botulism in humans.

When we swallow the spores (which we frequently do, since they are everywhere) no ill effects follow; the spores do not find the human intestinal tract fit for growth. However, in a sealed can or jar, the bacterium grows well, if the food is not too acid. Therefore, proper processing at adequately high temperatures is required to make processed foods safe. Commercially processed canned or smoked foods are carefully heat-treated to assure that the botulinum spores are killed, yet cases of botulism from such sources are on record.

The toxins commonly dealt with in this country—type A and type B—are very powerful. When toxic food is merely tasted, severe poisoning may result. Death is not infrequent in botulism.

Symptoms usually appear within 12–36 hours after the offending food has been consumed. They are: digestive disturbances, nausea, vomiting, sometimes diarrhea, headache, dizziness, and general malaise. Other important symptoms are double vision and difficulty in speaking, swallowing, and breathing. In the later stages of the illness the victim can not breathe sufficiently to sustain life and will die. Therefore, prompt medical aid must be sought at the onset of symptoms. Approximately two-thirds of all type A cases of botulism are fatal.

Food frequently involved in type A and B botulism are improperly canned items, especially vegetables. Other items, usually of a pH above 5, are also responsible although outbreaks of botulism involving food of pH below 5 (tomatoes, some fruits) are on record also. Botulism from type E toxin has been found primarily in connection with improperly processed fish products.

Storing canned items under hot conditions may allow spores of types A and B that have survived the canning process through faulty processing to germinate, multiply, and produce toxin. In contrast, spores of type E are able to germinate at very low temperatures, namely, at 38 F (3.3 C). Therefore, refrigerators used to store smoked and fresh fish should have temperatures below 38 F (3.3 C). Although by using proper smoking procedures type E spores should not survive, they may exist when processing is faulty; also, recontamination of the processed fish with spores has occurred when sanitary conditions in the processing plant were poor.

Danger signs in cans are bulging; broken seals; rusting; leakage; contents bubbly, spurting out on opening the can; off-color; and off-odor. These danger signs may range from very noticeable to quite obscure.

Warning: Do not try to find out about flavor, because you might become poisoned in the process. The poison *can* be rendered harmless by boiling the food for 15 minutes. Do not even feed suspect food to animals without having rendered it safe by boiling.

Caution: a canned food product containing the toxin will have a tendency to start bubbling as soon as heat is applied. This occurs because tiny bubbles of gas produced by the bacteria expand in heating and finally escape. This bubbling should not be confused with a true boil.

Principles of Control (Canned Items)

1. Use only commercially canned food in quantity food service.
2. Refuse donations of home-canned food; they may not have been processed according to the rules for safety from botulism.
3. Purchase canned food from a reliable vendor and store cans in a cool, dry place; use them within a year.

Danger signs in smoked fish may be absent. However, there may be signs of spoilage, such as off-color, off-odor, or change in texture, which would make one suspicious; however, there may not be any such signs.

Spores of the type E which may contaminate smoked fish, are easily killed, provided the smoking process is properly done. Therefore, smoked fish should be purchased from a reliable source through a reputable vendor. The spores of the type E can germinate at refrigerator temperature if given time.

Principles of Control (Smoked Fish)

1. Purchase smoked fish from a reputable dealer.
2. Use fish promptly, unless frozen. Keep unfrozen smoked fish under low-temperature, strict refrigeration; keep frozen fish solidly frozen.
3. Use frozen smoked fish promptly after it has thawed.

Note: For quick reference see Table IV in Appendix A.

STAPHYLOCOCCUS (STAPH) FOOD POISONING—A FOOD INTOXICATION

This type of food poisoning occurs very frequently. The cause is the toxin of *Staphylococcus aureus*. The organism is typical for all staphylococci in that the cocci bunch together in clusters, short chains, or pairs. "Aureus" means golden in latin, and this description fits the color of many of the strains when grown on solid culture media. The organism is facultative in its requirement for free oxygen; but in general it multiplies more profusely under aerobic than anaerobic conditions.

Various *Staphylococcus aureus* strains are capable of releasing a toxin into the food in which they can grow, and this toxin causes food poisoning.

Some of the toxin-forming (toxigenic) strains are very tolerant of high salt concentration and nitrites, which are frequently used in curing meat. They are also moderately tolerant of high sugar concentration. The toxic organism is capable of growing over a wide range of pH, about 4.5 to above 7; it can also grow over a wide temperature range, about 39 to 114 F (4 to 46 C), but growth and production of toxin are best at 70 to 98.7 F (21 to 37 C). It is obvious that this contaminant will find many environmental conditions suitable for its activities, and this must be taken into account when techniques for safe handling and holding of food are worked out. Staphylococci survive freezing and freezer storage.

Symptoms in people who have eaten toxin-containing food items appear on an average within 2–3 hours, but shorter and longer periods are possible.

The symptoms include salivation, nausea, vomiting, abdominal cramps, diarrhea, headaches, muscular cramp, sweating, chills, and prostration. Their duration is brief—from 1 to 2 days—and mortality is extremely low. In severe cases, medical treatment is required.

Source and Routes   The most important source of staphylococci capable of causing food poisoning is the human body (Figure 10). Even healthy persons carry these bacteria in their mouths, throats, and noses. Infected wounds are particularly rich in food-poisoning staph, and so are pimples, furuncles, infected throats, eyes, ears, sinuses and other pus-containing sores.

The organisms get into food if the food handler does not exercise care; for example, when the person coughs and sneezes over or near food or into the hands without washing them afterwards, or into a handkerchief without washing the hands afterwards; or when the nose is picked while preparing food (Figure 11), the face scratched, or pimple broken or when the person has an infected sore on the hands, even *if* he or she should try to protect it with a bandage. Foods frequently indicted in staphylococcal food poisoning are:

1. *Protein foods* and dishes made with them. Protein foods are meat, milk, poultry, eggs, fish and shellfish.
   The reason: Staph thrive in these items because of their composition and lack of acidity.

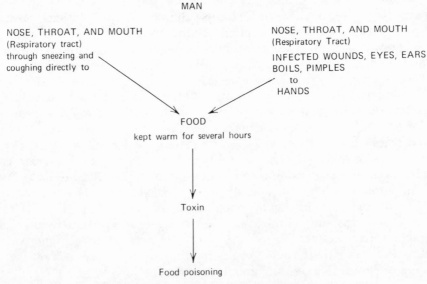

Figure 10. Sources and routes of staphylococci.

Figure 11. The nose is a rich source of staphylococci.

2. *Foods that are handled much,* as in grinding, slicing, deboning, shaping.
   The reason: The handling gives the staph a good opportunity to get into the food either directly from the food handler or indirectly from soiled tables and other working surfaces, soiled equipment and utensils, soiled clothing, and the like.
3. *Food that remains for hours in the temperature* DANGER ZONE *(45 to 140 F; 7.2 to 60 C) at which bacterial growth is possible.* This may happen when food stands around the kitchen during preparation hours; when it is forgotten overnight; when it cools slowly; and when it is held in hot-holding equipment that does not keep the food hot.

Caution: Food items contaminated with *Staphylococcus aureus* usually look, smell, and taste normal.

Can the staphylococci be killed by heating the food?
   We can kill staph in food that is heated to 165 F (74 C) or higher. Unfortunately, such heat treatment is impractical, or even impossible, for many menu items (see Part Three, Section G). Moreover, the toxins are

quite resistant to heating, especially type B toxin. Resistance to heat varies a great deal with a number of conditions. Therefore, hard-and-fast rules cannot be given for the destruction of the toxins once they are present in food. Preventing their formation by using the following methods is one of the most effective sanitary techniques to apply during every step of preparing menu items:

(a) Keep the food clean and prevent its contamination in the first place.
(b) Keep the food either hot, at least 140 F (60 C); or keep it cold, 45 F (7.2 C) and lower, to prevent multiplication of the staph and the formation of toxin.
(c) Cool food rapidly to safe temperatures.

Principles of Control

1. Persons suffering from respiratory ailments or pus-containing body sores should not handle food.
2. Food handlers should practice personal cleanliness.
3. Food handlers should have clean working habits and wash their hands after touching their body and everything else that is likely to be contaminated with staph.
4. Food should be manipulated with utensils rather than hands, whenever feasible.
5. Food preparation equipment and utensils must be maintained in a sanitary condition at all times.
6. Food should not be held at lukewarm temperatures for longer than a couple of hours. Formation of bacterial toxins must be prevented by keeping food either hot or cold. Staph toxin, once formed, is resistant to heat and can *not* be rendered harmless by heating the food.

Note: For quick reference see Appendix A, Table V.

SALMONELLA FOOD INFECTION OR SALMONELLOSIS

Salmonella food infections are very common. Many species and types (called serotypes) of salmonellae are known. The infection caused by salmonellae is called salmonellosis. Numerous living cells of the organism have to be present in the ingested food in order to produce illness.

The organism is a facultative aerobic rod, not capable of forming spores. It does not form toxin in the food in which it multiplies. Instead, the activity of the cells when they reach the intestinal tract of the victim causes an infection.

It should be mentioned that among the salmonellae is one species, *S. typhi,* a highly infectious pathogen capable of causing typhoid fever in humans. It is not regarded as a food-poisoning organism, since fairly few cells will suffice to make a person ill with the fever. Food may, however, serve as a vehicle of transmission.

The salmonellae grow at a wide range of acidity, from quite acid (pH near 4) to quite alkaline (pH near 9). The temperature range at which they may multiply is also wide. The lowest temperature that will support growth varies with the type of food and may be as low as 44 F (6.7 C), which is refrigerator temperature. The organism grows very well at room temperature, with 98.7 F (37 C) supporting growth best. At about 115 F (45.6 C) growth ceases. Salmonellae may survive freezing and storage in a freezer.

Symptoms in people who have eaten contaminated food items containing large numbers of *Salmonella* cells resemble those of *Staphylococcus* intoxication, but are slower to appear: the average is 12 to 24 hours, the range 3 to 72 hours. Mortality is low in victims. However, people who have had salmonellosis may become carriers. This means that they harbor the pathogens without outward signs of illness.

Sources and Routes   There are many sources (Figure 12); important ones are:

1. The intestinal tract of humans.
2. Human wastes—feces and sewage.
3. The intestinal tract of animals.
4. Animal wastes—feces and manure.
5. Food items already contaminated with salmonellae when delivered to the premises of the foodservice establishment; meat and poultry are examples of potential sources of salmonellae.

Source 1.   The intestinal tract of humans contains salmonellae when a person is ill with salmonellosis. Also, the intestinal tract of persons recovering from a bout with salmonellosis may harbor salmonellae for varying lengths of time; these persons are carriers. When an ill person or a carrier handles food, the salmonellae the person harbors may first contaminate the hands, which in turn contaminate the food. Hands may become contaminated with feces when a person visits the toilet. Thorough washing will remove a great number of contaminants. These are important questions:

Are hands thoroughly washed each time?
Are soap, plenty of hot water, paper towels or air dryers provided and used?

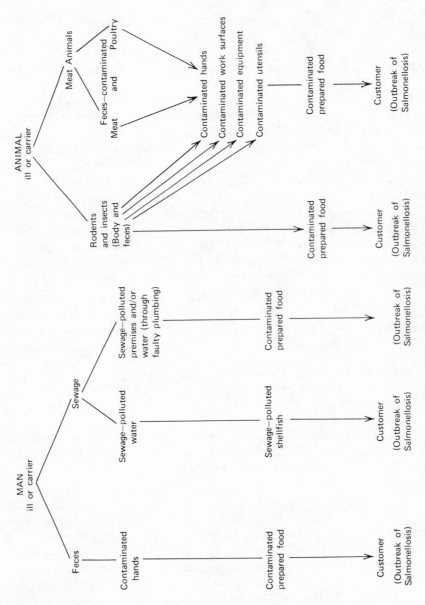

Figure 12. Sources and routes of salmonellae.

Removing contaminants effectively may be difficult when the salmonellae are present in great numers and/or when the skin of the hands is rough and the nails long. Therefore,

(a) *No person while ill with salmonellosis or while a carrier should handle food.*

(b) *Every food handler should thoroughly wash the hands after visiting the toilet.*

(c) *Adequate hand-washing facilities must be provided.*

(d) *The food handler should keep nails trimmed and clean.*

Source 2.   Sewage contains the bacteria present in the human intestinal tract. Sewage in turn may contaminate water and soil, and food harvested from contaminated water and soil. A good example is contaminated shellfish that has been harvested from sewage-polluted waters.

Source 3.   The intestinal tract of many animals contains salmonellae at one time or another. Among these animals are cattle, horses, hogs, poultry, dogs, cats, mice, rats, flies, and cockroaches.

Rodents, flies, and cockroaches contaminate food with their *Salmonella*-infested droppings. These animals feed on *Salmonella*-contaminated filth and carry salmonellae on their bodies.

Dogs and cats that may accidentally enter the food preparation area are a menace there. Pets clean themselves by licking, and their tongues become contaminated with salmonellae and other microorganisms of an undesirable nature.

Source 4.   Food contaminated with salmonellae when they are received at the food service premises can be an important source of these organisms in the kitchen. A great variety of foodstuffs have been found to be contaminated with salmonellae. Quite frequently contaminated are meat and meat products, and poultry and poultry products. When cattle, hogs, and poultry are slaughtered, the meat may become contaminated with salmonellae during processing.

In the past, processed eggs (such as frozen and dried eggs) have, at times, been found contaminated with salmonellae. This is being corrected by killing off the salmonellae through pasteurization of the egg mass before it is frozen or dried. Cracked shell eggs or severely cracked eggs may contain salmonellae. This is why buying "bargain" eggs is a very poor—in fact, dangerous—practice that is on the "No" list.

Shellfish harvested from polluted waters must be expected to be contaminated with salmonellae. Shellfish should come from safe sources and be certified by state agencies to be wholesome.

The salmonellae from a contaminated food supply may in turn contaminate prepared food items directly, or they may first contaminate hands, equipment, and utensils that then pass on the salmonellae to prepared food. Figure 12 depicts the direct and indirect sources of salmonellae.

There are a great variety of foods that have frequently been implicated in outbreaks of *Salmonella* foodborne infections. Common offenders are meats and poultry (turkey in particular) and the products made with them, such as meat and poultry pies and casseroles, meat and poultry sandwiches, turkey, stuffing, gravies; as well as egg custards, seafood harvested from sewage-polluted waters, and many other items.

The foods most frequently involved in outbreaks of *Salmonella* food infection are:

1. Meats and poultry.
   The reason: These foods are likely to be contaminated when purchased and are therefore especially hazardous.
2. Cracked eggs.
   The reason: The eggs contain salmonellae.
3. Menu items made with meat, milk, poultry, eggs, fish, and shellfish.
   The reason: The salmonellae thrive well in these items because of their composition and lack of acidity.
4. Foods that remained for hours in the temperature DANGER ZONE, 45 to 140 F (7.2 to 60 C), of bacterial growth.
   The reason: The salmonellae are able to multiply between 44 F (6.7 C) and 115 F (45.6 C).

Can the salmonellae be killed by heating the food?

We can kill salmonellae by heating contaminated food to 165 F (74 C) or higher. As is true for staphylococci, we cannot rely on this measure for all items except for certain specific cases to be discussed later on (see Part Three, Sections F and G). It is therefore necessary to prevent contamination with salmonellae during food preparation and to prevent those salmonellae that may be present in the food from multiplying by practicing strict time-temperature control during preparation, service, and storage. Keep food hot, 140 F (60 C) or higher; or keep it cold, 45 F (72 C) or lower; cooling hot food rapidly to safe temperatures.

Principles of Control

1. Persons suffering and/or recovering from salmonellosis, or who are carriers, must not handle food.
2. All food handlers must thoroughly wash their hands after visiting the toilet.
3. Food handlers must wash their hands after working with raw meats and poultry and before touching other food, particularly cooked items.

4. Food handlers should have separate work surfaces for raw and cooked foods, to avoid cross contamination.
5. Food handlers should prevent cross contamination. If the same work surfaces have to be used for raw and cooked foods, the surfaces must be thoroughly scrubbed and sanitized before they are used for cooked food. Remember this when disjointing raw meat or poultry on a work surface, then using the same surface for deboning and slicing the cooked meat.
6. Plumbing must be kept in excellent working order to prevent sewage from contaminating water and food.
7. Rodents and insects must be kept under strict control.
8. Pets must not be allowed in the areas where food is handled.
9. Meat and poultry should be government-inspected; shellfish should be certified to be wholesome; cracked eggs should not be used.
10. All equipment and utensils should be kept in an excellent condition of repair and cleanliness.
11. Excellent housekeeping should be practiced throughout the establishment.
12. Strict time-temperature control should be practiced during actual food preparation, hot-holding, and cooling Keep food hot, 140° F (60 C), or above; or keep it cold, 45 F (7.2 C), or below. Cool hot food rapidly, using efficient cooling techniques. For quick reference see Table VI in Appendix A.

PERFRINGENS FOOD POISONING

The gastrointestinal illness (gastroenteritis) caused by the causative organism, *Clostridium perfringens,* is considered by some bacteriologists to be an intoxication; by others, an infection. Although a toxin is formed by the pathogen, a large number of bacterial cells must be ingested to produce symptoms.

Clostridium perfringens (in the English literature it is referred to as *C. welchii*) is a spore-forming, anaerobic rod. The spores are very resistant to adverse conditions and are able to survive for long periods of time in soil. *C. perfringens* is a common inhabitant of the intestinal tract of humans and animals. When meat animals are slaughtered, the meat is likely to become contaminated with the organism. Therefore, we must expect its presence on raw meat and act accordingly, lest food poisoning might result. More will be said on the subject in Part Three.

Symptoms following the ingestion of food heavily contaminated with *C. perfringens* occur within 8 to 22 hours after the offending food item has been consumed. They are similar to symptoms typical of *Staphylococcus aureus* intoxication and *Salmonella* infection, but they are usually milder, and vomiting is usually absent.

The source of the organism is the intestinal tract of humans and animals (Figure 13). Consequently, it is also found in feces, sewage, manure, and soil. Kitchen dust is a rich source of the spores; in fact, the organism seems to be everywhere.

Since it is of intestinal origin, the organism has a good chance to contaminate meat after slaughter. The organism may grow over a wide range of acidity and alkalinity, with the pH range of 4 to 9 acceptable to it. Excellent growth takes place in foods that are near neutrality, namely, pH 6 to 7.5.

Although the temperature requirements of *C. perfringens* depend much on strain and growth medium (which includes many food items), the temperatures most favorable for growth are considered to be 109 to 116 F (43 to 47 C). Growth is scant at temperatures of 59 to 68 F (15 to 20 C) and at

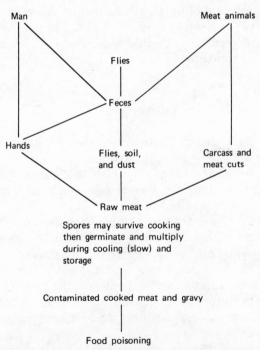

Figure 13. Sources and routes of *Clostridium perfringens*.

122 F (50 C). These temperatures are very important in connection with control measures involving all-important time-temperature control.

What foods are most frequently involved?

Most frequently indicted are meat and meat gravy; beef and veal are at the top of the list of offenders. Large cuts of rolled roasts are often implicated in perfringens food-poisoning outbreaks. The reason is that the heating in the innermost portion (especially when cooked to the rare stage) is not sufficient to kill perfringens spores, especially spores of heat-resistant strains. In fact, the heating might even stimulate the spores, making them ready for germination and subsequent production of a new crop of vegetative cells.

The following are examples of circumstances frequently associated with perfringens poisoning:

(a) Meat and gravy, also items such as stews and braised and boiled cuts, held warm for hours at temperatures *below* those recommended for hot-holding, which are 140 F (60 C) or higher.
(b) Meat was cooked in advance and cooled slowly, and, while cooling, the meat and gravy were actually in the DANGER ZONE for many hours.
(c) Gravy was held overnight on the back of the stove, neither hot nor cold.
(d) Cooked meat was recontaminated: meat was placed on table or board or other working surface on which raw meat had been handled previously and which was superficially cleaned or not cleaned at all. Or meat was sliced on a superficially cleaned slicer or with another unclean utensil.

Are the spores killed in routine cooking of meat?

Unfortunately, there are many strains of this bacterium around, and some are very resistant to heat. We cannot rely on heating to kill the pathogen; we must prevent its multiplication. To do this successfully, we need to know the conditions under which multiplication can occur.

1. Growth of *Clostridium perfringens* can take place between 60 and 122 F (15 and 50 C).
2. Growth is excellent between 109 and 116 F (43 to 47 C). These temperatures are dangerously close to those at which cooked meat is frequently held warm to prevent it from getting overdone before it is served.

Consequently, the most important control measures against perfringens food poisoning are (a) cleanliness and (b) strict time-temperature control.

Table 1. Some Less Common Types of Bacterial Food Poisoning

Illness	Cause	Symptoms	Foods Frequently Involved	Principles of Control
Vibrio parahaemolyticus infection	V. parahaemolyticus	Symptoms appear within 2 to 48 hours after contaminated food was eaten. They are: abdominal pain, diarrhea, nausea, vomiting, headache. Mild fever and chills may be present.	Raw saltwater fish and shellfish; crustacea; underprocessed fish products	Cook items thoroughly, avoid recontamination caused by contact with uncooked saltwater seafood through soiled surfaces and equipment, hands.
Escherichia coli Enteropathogenic infection	E. coli (2 types)	Symptoms appear within 8 to 44 hours, depending on type. They are: diarrhea for both types plus a number of other symptoms, depending on type.	A variety of food items	Cook food thoroughly, chill rapidly. Practice excellent personal hygiene and apply sanitary techniques in food preparation.
Bacillus cereus gastroenteritis (intoxication)	Toxin of B. cereus	Symptoms appear in from 1.5 to 16 hours and are similar to those of staphylococcal intoxication. They are: nausea, abdominal cramps, diarrhea, vomiting.	Predominantly cereal products, especially rice	Cool cooked items rapidly to safe temperatures.

| Shigellosis (bacillary dysentery) infection | *Shigella* (several species) | Symptoms may appear in 1 to 7 hours, and usually in less than 4 days. They are: headache, nausea, abdominal cramps, chills and fever, diarrhea, and general prostration. | A variety of food items have been indicted. | Personal hygiene, sanitary methods of food preparation, thorough heating of food. Rapid cooling. |
| Yersiniosis infection | *Yersinia enterocolitica* | Symptoms appear rather late, from 24 to 36 hours. They are: abdominal pain, diarrhea, nausea, vomiting, headache, and general malaise. | Meats (pork). A variety of contaminated raw or cooked food items. | Protect food from contamination. Cook food thoroughly. |

Principles of Control

1. Food handlers must prevent contamination of cooked food with *Clostridium perfringens* by using separate work surfaces for raw and cooked meats; if this is not feasible, they must thoroughly clean work surfaces after they have been used for raw meat.
2. Food handlers must wash their hands after working with raw meat and before handling cooked food.
3. The food-preparation area should be kept as free of dust as is feasible.
4. Meat slicers and other meat-cutting utensils must be properly cleaned after use.
5. Food handlers must thoroughly wash their hands after visiting the toilet and each time they pick up articles from the floor, lace their shoes, or handle crates, packages, and any other articles likely to be soiled.
6. Strict time-temperature control must be practiced in the preparation, hot-holding, and cooling of meat and gravy. Meat and gravy must be held hot (140 F, 60C, or higher) or cold (45 F, or 7.2 C, or lower); and cooling of hot food should proceed rapidly.

For brief reference, the characteristics of the *C. perfringens* organism are shown in Table VII in Appendix A.

Other less common types of food poisonings caused by bacteria and the principles of their control are presented in Table 1.

SECTION D

Foodborne Illness Caused by Agents Other than Microorganisms

Poisonous plants, poisonous animals, poisonous chemicals added to food, and metals from unsuitable cooking utensils, have all been reported as causes of food poisoning.

POISONOUS PLANTS AND ANIMALS

Certain mushrooms are poisonous, but mushrooms purchased from a reliable vendor are safe.

Green potatoes and sprouting potatoes have been reported to cause food poisoning. Refuse to accept such potatoes upon delivery.

Other plants that have been reported to cause food poisoning are not mentioned here, as these plants are not used in preparing food for the public. The interested reader is referred to Longrée, *Quantity Food Sanitation* (see Readings for Part Two).

Among animals that have been reported to cause paralytic shellfish poisoning are mussels and clams, from certain regions within the United States and Canada, that had fed on poisonous plankton (plankton are tiny ocean plants and animals that float in the water and serve as food to shellfish). However, seafood from approved waters is safe. The widespread notion that shellfish should not be eaten during the months *not* containing the letter "r" is false. It originated at a time of poor refrigeration during the warm months of the year.

A number of tropical fish are known to be poisonous and therefore inedible. But even wholesome fish may, at times, become poisonous when they feed on toxic materials. Ciguatera poisoning is common in certain areas of the Pacific and also throughout the Carribean; the toxin is heat stable. Fish may become poisonous through ingestion of poisonous plankton and also of mercury present in mercury-polluted waters. In quantity foodservice the use of poisonous fish is unlikely because seafood from approved sources must be used.

POISONING BY CHEMICALS

Chemical poisoning may be caused by negligent housekeeping, pesticides, and certain additives, or unduly large amounts of these.

Negligent Housekeeping

Frequent cases of food poisoning have been reported that were caused by disorder in the foodservice department—for example, when cleaning compounds, polishes, insecticides, and the like were stored together with food-preparation ingredients, or were not labeled, or both, and thus mistakenly used for food ingredients. There is no excuse for slipshod housekeeping.

Chemicals must be stored separately in closed, clearly labeled containers, away from the area where food is stored and prepared.

Additives

Chemical additives are being used—and have been for many years—in the preparation, processing, packaging, and storing of foods. They are used for various purposes: as an aid in enhancing the culinary quality of a food, such as flavor, texture, and color; and as a means of preserving food and extending its shelf life. Additives should never be used to cover up food spoilage.

A chemical that has become quite popular as a flavor enhancer of food prepared in the home and in quantity as well is monosodium glutamate (MSG). It can act as a poison when used in large amounts in a menu item. MSG is a popular ingredient in Chinese menu items, and food poisonings are on record that were traced to Chinese restaurants and overzealous use of MSG. The symptoms can be dramatic, involving headache, nausea, and flushing of the face.

The use of chemical additives in processed foods is regulated by the Food and Drug Administration (FDA). Safety policies for the use of additives are established with the aim of protecting the consumer from those that might cause illness in humans. The difficulties encountered in establishing that an additive is safe, and at what levels, can be formidable.

Enactment of the laws regulating the use of food additives dates back to the time (1958) when appropriate amendments were added to the Federal Food, Drug and Cosmetic Act of 1938. In general, the amendments require the proponents of a new food additive must prove its safety and effectiveness and that they must obtain permission for its use in food from the Food and Drug Administration. Exempted from the requirement of proving the safety of an additive are those chemicals that over many years have been generally recognized as being safe; these make up the so-called GRAS ("generally recognized as safe") list. This list is periodically re-evaluated, because increasingly more sensitive methodology for testing safety tends to eliminate additives previously considered acceptable and safe.

The much-debated Delaney Amendment prohibits the use of any additive if "found to induce cancer when ingested by man or animal; or if it is found, after tests which are appropriate for the evaluation of the safety of food additives to induce cancer in man or animal."[*]

Before going further, it now seems advisable to define the term "food additive" as given in the Federal Food, Drug and Cosmetic Act, which says:

The term 'food additive' means any substance the intended use of which results or may reasonably be expected to result, directly or indirectly, in its becoming a component or otherwise affecting the characteristics of any food (including any substance intended for use in producing, manufacturing, packing, processing, preparing, treating, packaging, transporting, or holding food; and including any source of radiation intended for any such use), if such substance is not generally recognized, among experts qualified by scientific training and experience to evaluate its safety, as having been adequately shown through scientific procedures (or, in the case of a substance used in food prior to January 1, 1958, through either scientific procedures or experience based on common use in food) to be safe under the conditions of its intended use; except that such term does not include—
(1) a pesticide chemical in or on a raw agricultural commodity; or
(2) a pesticide chemical to the extent that it is intended for use or is used in the production, storage, or transportation of any raw agricultural commodity; or
(3) a color additive; or
(4) any substance used in accordance with a sanction or approval granted prior to the enactment of this paragraph pursuant to this Act, the Poultry Products Inspection Act . . . or the Meat Inspection Act . . . ; or
(5) a new animal drug.[†]

Antimicrobial Preservatives   These are additives used to enhance the microbiological quality of the foods into or onto which they are added. They are supposed to kill—or at least inhibit the growth of—microorganisms that may be bacteria, molds, or yeasts. These microorganisms might be spoilage organisms or pathogens. The additives should do what they are expected to do in amounts that are safe for human consumption. They should not adversely affect the culinary quality of the food for which they are used and they should be economical as well.

---

[*]*Federal Food, Drug and Cosmetic Act as Amended October 1976.* Sec. 409(c)(3)(A). Washington, D.C.: Government Printing Office.
[†]*Ibid.*, Sec. 201(s).

Among the substances used as preservatives are old and new ones, organic and inorganic ones. Some quite familiar old timers include acetic, lactic, and citric acids and their salts. Other common preservatives are sugar, salt (sodium chloride), alcohols, spices, and substances present in wood smoke.

Vinegar (acetic acid) is used largely in cucumber and similar pickles as well as in ketchup and other such sauces. Vinegar is also an ingredient in mayonnaise and pickling solutions used for some meats, such as pigs' feet, and in pickled herring. Acetic acid is more effective against bacteria and yeasts than against molds.

Sugar in high concentrations is an essential preservative in products made from fruit such as jams, jellies, and preserves. Sugar is also an important ingredient in sweetened condensed milk.

Salt is an important component of brines used for curing meats.

Both salt and sugar tie up water, thereby making essential moisture less available to microorganisms.

Certain ingredients present in *wood smoke* contribute to the keeping quality of smoked meat, poultry, and fish.

Examples of other organic acids and/or their salts that are useful in extending the shelflife of certain foods are:

Propionates used in the prevention of molds in baked products and cheese spreads.

Benzoates, the salts of benzoic acid used in numerous items, among them relishes, fruit juices, jellies, jams, and margarine; they are particularly effective against molds.

Sorbates, which are applied either as a spray or dip, or incorporated into a food; sorbates are used extensively in baked goods, cheese products, jellies, jams, and on dried fruit. They are also utilized as coatings for packaging materials.

Sulfur Dioxide and Sulfites   These chemicals are employed extensively (in aqueous solutions) as sanitizers for equipment used in the wine industry. Sulfur compounds are also effectual in treating some dehydrated fruits, and on vegetables to be dehydrated.

Nitrites   These have been used for many years in the curing of meats, poultry, and fish. They improve the color and flavor of the cured products and also inhibit the growth of *Clostridium botulinum,* cause of deadly botulism; and of *Clostridium perfringens,* a bacterium capable of causing perfringens food poisoning. The use of nitrites has been heatedly debated in the recent past by scientists, food processors, and regulatory agencies, because nitrites may possibly be dangerous as a result of their potential for causing cancer. Nitrite exposed to high cooking temperatures, as when cooking bacon, may form cancer-forming compounds, nitrosamines. Another concern about the safety of nitrite has arisen from laboratory research with test animals that also indicated a cancer-forming potential. More tests are needed to shed light on this very important problem. In the meantime, a reduction in the amount of nitrite used in curing food eases the situation. By law, nitrosamine in bacon is now reduced to a nonconfirmable level. The USDA has set the confirmable level of nitrosamine as 10 parts per billion. Because of the reduced shelflife of bacon, freezer-storage is advocated.

Other additives are being tested that in the future may take the place of nitrites in the curing of meat, poultry, and fish. It is therefore possible that the use of nitrites will be gradually phased out as safe and feasible substitutes are made available to the processors.

Pesticides

Pesticides are supposed to be applied to growing food plants according to schedules that should ensure that the chemicals do not adhere to the food in injurious amounts at harvesting time. However, traces may remain; all fruit and vegetables must be carefully washed before they are used. Pesticides may also be taken up by food plants and become part of them as the chemicals are incorporated into their cells.

The Environmental Protection Agency sets tolerances for the amounts of pesticides that may remain on, or in, food plants. The Food and Drug Administration enforces the established tolerances.

Animal Drugs

Hormones For a number of years a growth-promoting hormone, diethylstilbestrol (DES), had been used in food animals. This hormone has shown potential to cause cancer. Since it can not be assured that residues of DES have completely disappeared from the flesh of food animals, even if food containing the hormone is withdrawn within a reasonable time prior to slaughtering, the use of DES is now being banned.

Antibiotics The distribution and use of animal feed containing penicillin, chlortetracycline, and oxytetracycline is subject to special approval from the Food and Drug Administration. Antibiotics are no longer allowed to be used as dips for poultry and fish. The reason underlying the ruling against the use of these antibiotics (except by special permission) is the dangerous possibility that bacteria, including pathogens such as salmonellae, which exist in or on the animal destined for use as meat for humans, might develop a resistance toward these antibiotics. If such resistant pathogens should cause illness in humans, the antibiotics administered to the patients might not be effective.

Metals

Metals may get into food by the use of food-preparation and storage utensils unsuitable for this purpose. The following metals have been reported to cause food poisoning: cadmium, antimony, lead, zinc, and occasionally copper.

Cadmium, when employed as plating material for ice-cube trays, refrigerator shelves, and syrup pumps, has been reported to cause poisoning. Antimony, which occurs as an impurity in grey enamelware, and zinc, which is used in galvanized containers, may render acid food poisonous. Copper has been reported to cause poisoning when in prolonged contact with acid foods and carbonated beverages.

Containers, shelves, and pipes containing the above materials may present hazards from poisoning, especially when used with acid food and drink.

Lead is condemned for plating purposes. When used in solder, lead must be kept at a low percentage. The health authorities should be consulted about applicable regulations.

Mercury, copper, lead, and zinc have been shown to, at times, contaminate seafood. Government agencies are concerned with surveying fishing waters for pollution with metals and are imposing fishing restrictions where necessary.

SECTION E

Reporting and Investigating Foodborne Illnesses

One important means of preventing future incidences in outbreaks of foodborne illness is to learn from the mistakes of the past. Therefore, we must trace all incidences to their causes. The foodservice manager does not possess the technical know-how that a sanitarian and a bacteriologist have; therefore, he or she needs the help of the health authorities in tracing the cause of the incident and in correcting the situation to prevent reoccurrence.

Reporting to the health authorities a food-poisoning outbreak, or the suspicion of one, should be prompt. Delay in reporting not only severely hampers the investigation, but may also allow the illness to spread. In some states reporting is required by law; in others it is voluntary. Consult your local health authorities.

Whether reporting is compulsory or not, it is strongly recommended that all suspected cases of foodborne illness be reported to the health authorities; they can be of tremendous help to management in preventing future incidences.

An analysis of a 5-year period of reporting has shown that approximately 37 percent of the reported outbreaks took place in foodservice establishments and only 14 percent in homes; and that even fewer were traced to food-processing plants. The remaining outbreaks could not be properly traced.

Outbreaks of food poisoning are reported by the local health authorities to those of the individual states which in turn report to the Public Health Service, Center for Disease Control (in Atlanta, Georgia). This agency publishes reports on food poisoning incidences in a weekly, *Morbidity and Mortality Weekly Report,* and also prepares an *Annual Summary* (both available upon request).

The investigation of an outbreak of foodborne illness seeks to accomplish these tasks:

(a) Find which meal and specific menu item caused the illness.
(b) Determine the organism or other agent that caused the outbreak.

(c) Determine where and how the food became contaminated.

(d) Determine the opportunity for bacterial multiplication in the food.

To do this, a team of investigators will question the victims, and visit the foodservice establishment to interview management and employees, take samples of suspected menu items if they are still available, and check on the sanitary condition of the operation.

Since many food-poisoning organisms are harbored in the human body, the persons who prepared the suspected meal—or menu items—are examined to detect acute or recent illness and the presence of carriers. Specimens may be taken from food handlers to detect the possible presence of food-poisoning bacteria in and on their bodies.

Swabs are also taken from surfaces of tables, cutting blocks, and other food-preparation equipment to determine whether they are contaminated, and with what organisms or poison.

Food samples, specimens taken from food handlers, and swabs taken from equipment, are handed over to a laboratory for examination. All information is organized in a report.

It is extremely important that management and every employee interviewed cooperate to the fullest with the health authorities in their endeavor to clarify the causes of the incident. The health authorities will have valuable suggestions to make as to what needs to be done to remedy the situation and to avoid a repetition of similar incidents in the future.

What have past investigations taught us?

Of the outbreaks reported, very few are waterborne and milkborne, while over 90 percent are foodborne (Figure 14). Among the outbreaks caused by food, the chief offenders are beef (in particular, roasts and gravy), ham, and poultry and items made with them. Cream-filled bakery products and fish and other seafood are also responsible.

Staphylococci, salmonellae, and *Clostridium perfringens* are microorganisms frequently cited as causes of food-poisoning outbreaks. For many outbreaks the cause remains, unfortunately, unknown and many incidents are never reported.

The following circumstances are most frequently associated with outbreaks: mishandling of food through contact with unclean or ill food handlers, food handlers untrained or poorly trained in sanitary techniques; uncleaned, unsanitized equipment, utensils, and other food-contact surfaces; and, finally, long exposure of food to warm temperatures favorable to bacterial growth through holding food at kitchen temperatures, cooling food in batches too large for rapid chilling to safe temperatures; and long holding on serving counters on which the food was neither sufficiently hot or cold to preclude bacterial activity.

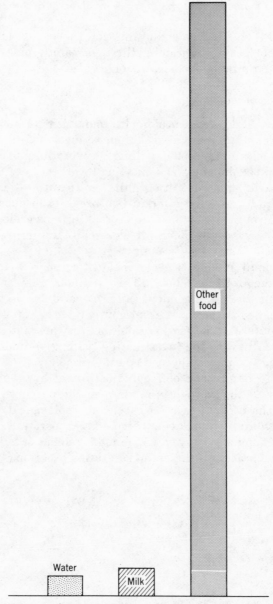

Figure 14. Comparative incidence of foodborne illness caused by water, milk, and other food.

What types of food-service establishments are involved in outbreaks of foodborne illnesses?

The answer is, all kinds. Restaurants top the listing of reported outbreaks, but schools, colleges, social gatherings, camps, private clubs, nursing homes, and hospitals have been involved also.

Recommendations for the manager to follow:

1. Invite the local department of health to participate in your sanitation program. Solicit their advice; ask them to speak before your employees on matters of sanitary food handling. Do everything to create a feeling of "being in this business of sanitation together."
2. Seek the advice and cooperation of the health authorities in all matters of health and sanitation *before* an outbreak of foodborne illness has occurred in your establishment. Outbreaks can be prevented by appropriate sanitary food management. The health authorities will help you to assess your operation for sanitation.
3. Retain a small sample of every menu item served each day, under refrigeration, for 24 to 48 hours. Then, if an outbreak of food poisoning should occur, the establishment can prove or disprove the charge.
4. Notify the health authorities promptly in case of an outbreak or suspected outbreak and save a sample of every menu item for analysis.
5. Cooperate fully with the investigating team; give information freely, and stimulate your employees to do the same.
6. Prevent the occurrence of foodborne illness by applying sanitary techniques all the way—beginning with the purchase of a safe food supply to the protection of food from microbiological, chemical, and physical contamination at every step: in transport and storage, during preparation and service, and in the management of leftovers.
7. Develop and practice a successful sanitation program (see Part Four).

Important Facts in Brief for Part Two

- Spoilage commonly refers to food decomposition and other damage that interferes with food quality and renders food unfit to eat.
- Spoiled food does not necessarily present a hazard from a public-health point of view, but spoiled food should be regarded as suspect. Spoilage may cause heavy financial losses to the foodservice business. Spoilage indicates poor management.
- Disease transmission refers to the fact that germs causing communicable diseases may be passed on from one person to others by way of food.
- Food transmitting disease germs never appears spoiled.
- Transmission may be direct from person to food to person; or it may be indirect. Important in-between vehicles of transmission are sewage, sewage-polluted water, rodents, and insects.
- The source of the human pathogens that are frequently transmitted through food is a person's body, in particular the respiratory and intestinal tract.
- The persons most likely to transmit disease-producing germs to food and eventually to the customer are the food handlers; they are the people who prepare the food, serve it, and clean the food-preparation and eating utensils. Therefore, those who are ill with communicable diseases or are carriers are a menace in the foodservice establishment.
- Carriers can be particularly dangerous because they do not look or feel ill and are likely to report to work. Carriers are known to have infected large numbers of customers before being detected as carriers.
- Certain germs causing animal diseases can be transmitted through food to humans and cause disease in people.
- Bacterial food intoxications and food infections are caused by consuming food containing bacterial toxins or high numbers of certain infectious bacteria, respectively. Symptoms following the consumption of the offending food are, usually, violent nausea, vomiting, abdominal cramps,

diarrhea, and general weakness. Foods become poisonous or infectious because they are mishandled. They usually look, smell, and taste like wholesome food and do not give the appearance of being spoiled. Organisms most frequently responsible for food poisoning and food infections are staphylococci, salmonellae, and perfringens bacteria.

Most of the organisms come originally from the body of humans and animals. With proper sanitary care their entry into food can and must be controlled at all stages of food preparations; and their multiplication in food can and must be prevented through strict control of time and temperature.

Here are the rules: Keep food clean, and either hot or cold. Cool warm food quickly to safe temperatures. Application of these rules requires the cooperation of everyone in the establishment. Sanitation is everyone's responsibility every minute of the day and every day of the year.

General control measures to prevent food spoilage, disease transmission and food poisoning are:

1. Provide and maintain an adequate, clean facility: clean kitchen, dry storage, refrigerators, and freezers; clean locker rooms and toilets; clean garbage disposal; and effective and safe rodent and insect control.
2. Provide and maintain adequate, sanitary food-preparation equipment and utensils and sanitary eating and drinking utensils.
3. Be informed about, and practice, the so-called OSHA regulations that deal with safety and health protection of your workers. OSHA stands for "Occupational Safety and Health Act." The Act became effective in 1971, and is primarily administered by the U.S. Department of Labor.
4. Purchase a sound food supply from reliable sources and use sanitary storage procedures.
5. Employ healthy food handlers and continuously train them to be health-conscious, personally clean, and eager to maintain good sanitary standards in their work.
6. Use sanitary procedures aimed at preventing bacterial contamination of food during storage, actual preparation, holding, and service. Maintain effective control of time and temperature at every stage of preparation, hot-holding, cooling, and serving of the food.
7. Be convinced that sanitation is important; set high sanitary standards, emphasizing their importance to employees and seeing to it that these standards are constantly maintained.

Detailed control measures will be discussed in Part Three.

Study Questions for Part Two

1. Define spoilage of fresh foods.
2. What are signs of spoilage of canned foods? Explain.
3. State the difference between disease transmission and food poisoning.
4. Name some very important diseases that have been known to be transmitted through food.
5. What are some important routes by which pathogens are transmitted from people to prepared food? Explain.
6. What is a carrier and what is the carrier's role in disease transmission? Explain.
7. Explain the meaning of food intoxication and food infection. Give examples of each.
8. Define potentially hazardous foods.
9. Discuss the basis for applying strict temperature control in the preparation and serving of menu items.

Readings for Part Two

Bryan, F. L. *Diseases Transmitted by Foods*. U.S. Dept. HEW Publ. No. (CDC) 76-8237, Center for Disease Control, Atlanta, Ga.

Center for Disease Control. *Foodborne and Waterborne Disease Outbreaks. Annual Summary*. U.S. Dept. Health, Education and Welfare,* Publ. Health Service, Atlanta, Ga. (Published on an annual basis).

Center for Disease Control. *Morbidity and Mortality Weekly Report*. U.S. Dept. Health, Education and Welfare, Publ. Health Service, Atlanta, Ga.

Christie, A. B. and M. C. Christie. *Food Hygiene and Food Hazards*. 2nd ed. Faber and Faber, London. 1977.

Community Nutrition Institute. *Nitrite: Why The Debate*. 20 pages. Available by request and payment of 50¢, from Consumer Division, Community Nutrition Institute, 1146 19th Street N.W., Washington, D.C. 20036.

Food and Drug Administration. "Carcinogenic Residues in Food Production Animals." *Federal Register, 44* (March 20): 17071. 1979.

Frazier, W. C. and D. C. Westhoff. *Food Microbiology*. 3d ed. McGraw-Hill, New York, 1978.

Furia, T. E. *Handbook of Food Additives*. The CRC Press, Cleveland, 1972.

Hobbs, B. C. *Food Poisoning and Food Hygiene*. 3d ed. Edward Arnold, Ltd., London. 1974.

Longrée, Karla. *Quantity Food Sanitation*. 3d ed. Wiley-Interscience, New York, 1980. Chapters I, III, IV, V, VI, VII.

*Now: U.S. Dept. Health and Human Services.

National Academy of Sciences. *Toxicants Occurring Naturally in Foods.* 2d ed. Washington, D.C. 1973.

Nickerson, J. T. and A. J. Sinskey. *Microbiology of Foods and Food Processing.* American Elsevier, New York. 1972.

Riemann, H. and F. L. Bryan, editors. *Food-Borne Infections and Intoxications.* 2nd ed. Academic Press, New York. 1979.

Skinner, F. A. and J. G. Carr, editors. *Microbiology in Agriculture, Fisheries and Food.* Academic Press, New York. 1976.

U.S Department of Health, Education and Welfare, Public Health Service, Food and Drug Administration. *Food Service Sanitation Manual.* 1976 Recommendations. DHEW Publ. No. (FDA) 78-2081. 1978.

Zottola, E. A. *Botulism.* Univ. of Minnesota, Agric. Ext. Service, Ext. Bul. 372, St. Paul, Minn. Revised 1976.

Zottola, E. A. *Clostridium perfringens food poisoning.* Univ. of Minnesota, Agric. Ext. Service, Ext. Bul. 365, St. Paul, Minn. Revised 1979.

Zottola, E. A. *Salmonellosis.* Univ. of Minnesota, Agric. Ext. Service, Ext. Bul. 339, St. Paul, Minn. Revised 1980.

PART THREE

Sanitary Practice

Clean Facility (Physical Plant)

The physical plant of a foodservice operation refers to:

1. Kitchens and all other areas where food is prepared, stored, and served.
2. Areas where equipment is washed and stored.
3. The dressing or locker rooms for employees, and toilet rooms.
4. The area for garbage disposal.

Sanitation of the physical plant involves:

1. Cleanability of walls, ceilings, and floors as a result of good construction, smooth materials, and good state of repair.
2. Proper sewage disposal and plumbing; presence of floor drains where needed because of spillage or method of cleaning floor.
3. Good ventilation.
4. Good lighting.
5. Adequate hand-washing facilities.
6. Sanitary dressing rooms or locker rooms and sanitary toilet facilities.
7. Sanitary garbage disposal.
8. Control of rodents and insects.
9. Separate storage area for cleaning materials and utensils.
10. Regular and adequate cleaning of facility.

CLEANABILITY OF WALLS, CEILINGS, AND FLOORS

Walls and Ceilings

Walls and ceilings should be light colored, smooth, nonabsorbent, and easily cleanable. Concrete or pumice blocks used for interior wall construction should be finished and sealed to provide an easily cleanable surface.

Exposed utility service lines and pipes should not obstruct or prevent cleaning of walls and ceilings.

Floors

Floors and floor coverings should be constructed of smooth durable material. Antislip floor covering may be used in areas where necessary for safety reasons. Carpeting is prohibited in food preparation, equipment-washing, and utensil-washing areas where it would be exposed to large amounts of grease and water, in food storage areas, and toilet areas where urinals or toilet fixtures are located. Mats and duckboards should be of nonabsorbent, grease-resistant materials and of such design and construction that they can be easily cleaned. Exposed utility service lines and pipes should be installed in a way that does not obstruct or prevent cleaning the floor.

Walls, ceilings, and floors must be maintained in a good state of repair since surfaces in poor repair are impossible to keep clean. Rough surfaces and surfaces with rough and open places collect debris, grime, and dust and provide harboring places for vermin, such as cockroaches. Cracks may provide entry places for rats and mice. Therefore, cracks must be sealed and rough surfaces changed to smooth, cleanable ones.

ADEQUATE SEWAGE DISPOSAL AND PLUMBING

Sewage Disposal

Improper sewage disposal can cause, and has caused, serious illness and outbreaks of food poisoning. Sewage contains dangerous germs, and must never be allowed to contaminate food or equipment and the area where food is stored or prepared.

Improper sewage disposal can:

1. Pollute the water.
2. Pollute food through overhead leakage of pipes.
3. Contaminate equipment.
4. Attract flies and other insects; these insects, in turn, contaminate food.

Sewage must be disposed of in a manner satisfactory to the health authorities, which means that sewage must be directed into:

1. A public sewage system.

2. A sewage system that is constructed and operated according to the rules and regulations of the health authorities; consult these.

Plumbing

Improper plumbing is a hazard for these reasons:

1. Cross connections between pipes carrying drinking water and pipes carrying nondrinking water.
2. Drainage stoppage.
3. Overhead leakage.
4. Backflow of sewage into drains from refrigerators, drains, dishwasher drains, etc.

All plumbing must be installed and maintained in a manner satisfactory to the health authorities. Plumbing must be maintained in good working condition, and any malfunction, such as leakage, stoppage, and flooding, should be reported by employees to the supervisor or manager so that the hazard can be corrected at once. The piping of any nonpotable water system must be durably identified so it is readily distinguishable from piping that carries potable water. The potable water system must be installed to preclude the possibility of backflow. A hose must not be attached to a faucet unless a blackflow prevention device is installed.

Floor Drains

Properly installed, trapped floor drains should be provided in all areas where water is spilled on the floor during normal operations or where floors are cleaned by hosing. State, city, and county health departments may have special regulations regarding floor drains. Consult your local health authorities.

Floor drains must be kept in good operating condition. When stoppage occurs, the supervisor or manager should be informed and the condition corrected at once.

VENTILATION

Kitchens, storerooms, dishwashing rooms, dining rooms, locker or dressing rooms, toilets, and storage rooms for garbage and rubbish should be well ventilated. Ventilation systems should be installed and maintained in a manner satifactory to the health authority. Examples of ventilation systems are window exhaust fans, hood exhaust systems, unit blowers, and unit cooling systems.

What proper ventilation can do:

1. Chase out air that has become hot, stale, odorous, moist, greasy, smoky, and the like, because of cooking, dishwashing, and various other causes.
2. Prevent deposit of droplets of moisture on walls and ceilings. Moisture furthers growth of microorganisms. Droplets containing bacteria and mold may fall onto food and food-preparation surfaces and equipment and contaminate them.
3. Minimize the soiling of walls, ceilings, and floors.
4. Facilitate removing air contaminated during cleaning or food preparation processes, or as the result of exhaust from other equipment.

LIGHTING

Good lighting in particularly needed in these areas:

1. Kitchens—in order to prepare food in a sanitary manner.
2. Food storage rooms—in order to recognize the condition of the food, to avoid spills, and to facilitate cleaning.
3. Areas where pots, pans and utensils are washed—to see that these get clean.
4. Areas where hands are washed—to see that hands and nails get clean.
5. Dressing or locker rooms—to see that the whole person looks tidy and clean.
6. Toilets—to see that they are in a sanitary condition and to facilitate cleaning.
7. Storage areas for rubbish and garbage—to check on orderliness, cleanliness, and freedom from vermin.

With good lighting, dirt becomes more conspicuous. With poor lighting, good cleaning is impossible.

Shielded light fixtures should be provided to protect food, utensils, and equipment from glass fragments in case the fixture should break. The shielding also prevents many instances of breakage and provides a greater degree of on-the-job safety to employees.

ADEQUATE HAND-WASHING FACILITIES

Responsibility of Management

1. Hand-washing facilities should be made available in several locations. Several states have legal specifications regarding hand-washing facilities. Since hands are the most important sources of germs with which food becomes contaminated, management should provide hand-washing facilities in locations where hands are likely to become contaminated with dangerous germs:
   (a) Adjacent to toilet rooms.
   (b) In locker or dressing rooms.
   (c) In areas where food is prepared.

No employee will walk great distances to wash hands every time they should be washed. Therefore, hand-washing facilities should be placed conveniently and be made available in adequate numbers. They should be installed so as to satisfy the requirements of the health authority. Sinks used for food preparations or for washing equipment or utensils should not be used for hand washing.

2. Hand-washing facilities should consist of:
   (a) A wash bowl equipped with hot and cold (or tempered) water.
   (b) Liquid or powdered soap or detergent.
   (c) Individual towels or other hand-drying devices such as air dryers.
3. Maintenance should be meticulous. The wash bowls, dispensers for soap, and other drying devices should be regularly and thoroughly cleaned. Soap and towels should be checked and replenished before they run out.

DRESSING OR LOCKER ROOMS AND TOILET FACILITIES

Responsibility of Management

1. *Dressing rooms or locker rooms* should be available so foodservice personnel may change their clothes and store their belongings. Street clothes can be a source of undesirable bacteria and should not be worn or hung in the kitchen. Also, clothes should not be hung in storerooms or toilets. Dressing or locker rooms should be located outside the area where food is prepared, stored, and served. Uniforms should not be worn outside the foodservice establishment. Dressing or locker rooms should be well screened and provided with proper hand-washing facilities. Mirrors should be hung away from hand-washing facilities.

Dressing or locker rooms should be kept in a sanitary condition at all times. They must be scheduled for regular and thorough cleaning. Floors and washbowls should be cleaned daily.

2. *Toilet facilities* should be provided in areas separate from those where food is prepared, stored, and served. Toilet rooms should be well screened and have tight-fitting and self-closing doors. Toilet rooms and facilities (such as commodes and urinals) should be kept clean and in good repair. Toilet tissue should be available at all times. Receptacles should be available for waste materials and should be emptied frequently, at least once a day.
3. Hand-washing facilities should be available in or next to dressing rooms and toilet rooms should include washbowls, a hot and cold or tempered water supply, soap or detergent, and towels or air dryers. Any self-closing or metering faucet used should be designed to provide a flow of water for at least 15 seconds without the need to reactivate the faucet. Steam-mixing valves are prohibited.

It is of utmost importance that everyone wash their hands after visiting the toilet. Signs should be posted as a reminder. The Health Department will furnish the signs upon request.

Washrooms must be immaculately maintained: floors, washbowls and mirrors cleaned daily; and soap and towels checked and replenished if necessary. SOAP AND TOWELS MUST BE AVAILABLE AT ALL TIMES.

The responsibilities of employees are to:

1. Put your clothes away neatly.
2. Don't leave articles around or clutter up space that ought to be available to everyone.
3. Don't comb your hair over the washbowl, creating a disgusting mess.
4. Keep toilet seats clean.
5. Thoroughly wash your hands after dressing and after visiting the toilet.
6. Report to your supervisor any disfunctioning of equipment or lack of toilet tissue, soap, and towels.
7. Cooperate to the fullest with fellow employees and management to keep areas neat and clean. Clean up after yourself and leave these places in a condition in which you would hope to find them yourself.

It is the responsibility of everyone to keep dressing or locker rooms, toilets, and washrooms tidy and clean.

SANITARY GARBAGE DISPOSAL

Improperly disposed garbage is a nuisance as well as a source of danger to sanitation. Unpleasant odors are a nuisance. The real dangers of *improperly* stored garbage are these:

(a) It becomes a harboring place for vermin such as rats, mice, flies and cockroaches.
(b) It represents a source of contamination of food and of equipment and utensils used in food preparation.
(c) It makes good housekeeping a difficult, if not impossible, task.

The responsibilities of management are to:

1. Provide garbage containers of durable, easily cleanable insect-proof and rodent-proof materials that do not leak and do not absorb liquids. Plastic bags and wet-strength paper bags may be used to line these containers. A sufficient number of containers should be provided to hold all the garbage and other refuse between collections. Cans kept outside should sit on a rack (Figure 15).
2. Provide either tight-fitting covers for the cans or a special, enclosed, vermin-proof, cool garbage room. A refrigerated garbage room aids in preventing rapid decay of garbage and the development of bad odors resulting from decay.
3. After cans are emptied, have them thoroughly cleaned inside and out; provide suitable facilities, including hot water or steam, detergent, and brushes for the job. In large operations, steam cleaning devices or can-washing machines may be available.
4. Have the storage area for garbage cleaned regularly and frequently.
5. Check frequently for signs of vermin activity.

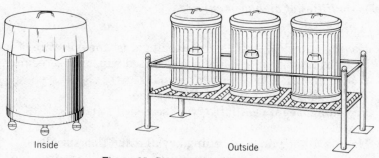

Inside   Outside

Figure 15. Sanitary disposal of refuse.

6. Install mechanical disposers (Figure 16) to reduce the amount of garbage and the number of unsightly garbage cans. If used, these mechanical disposers should be installed and maintained according to law. Check with your health authority.
7. Dumpsters, compactors, and compactor systems should be easily cleanable, provided with tight-fitting lids, doors, or covers and kept covered when not in actual use. Liquid waste from compacting or cleaning operations should be disposed of as sewage.
8. Dispose of garbage often enough to prevent the development of odor and the attraction of insects and rodents.

The responsibilities of employees are to:

1. Avoid spilling garbage and clean up spills if they happen.
2. Replace lids on cans after these have been opened. Containers used in food preparation and utensil washing areas should be kept covered after they are filled.
3. Carefully close doors of the garbage rooms to shut out vermin.
4. Report to supervisors any signs of vermin activity.
5. Do a thorough job of cleaning the cans and the storage area for garbage.

CONTROL OF RODENTS AND INSECTS

The presence of rodents such as rats and mice, and insects such as flies and cockroaches, is a menace to a foodservice operation. Rodents and insects may transmit disease by contamining food and food-preparation equipment and utensils. Their entry into the establishment must be prevented (Figure 17), and specimens present must be killed.

The responsibiliities of management are:

1. *To keep out flies by:*
   (a) Having outside openings protected against the entrance of insects by tight-fitting, self-closing doors, closed windows, screening, controlled air currents, or other means.
   (b) Providing screened storage for garbage.
   (c) Providing clean, ventilated, well-screened toilets.
2. *To control cockroaches by:*
   (a) Having all openings in ceilings, walls, and floors sealed.
   (b) Having damp areas eliminated, such as overflows, drippings from drains, and leaks from pipes; and crevices around equipment and pipes.

Figure 16. Mechanical waste disposer. (Courtesy of Hobart Mfg. Co.)

(c) Having all incoming boxes of foodstuffs inspected for presence of roaches and refuse roach-infested goods.

(d) Inspecting kitchens, storage rooms, and garbage rooms frequently for the presence of vermin. Remember that cockroaches are night insects that come out of hiding at darkness. Signs of cockroaches are their droppings, which, when dry, look like mouse pellets but possess lengthwise ridges and a musty odor.

(e) Hiring a professional exterminator for regular chemical control of vermin.

(f) Having the premises cleaned every night so that no crumbs and bits of food are left as food for vermin.

(g) Having the premises inspected every night for left-out food.

3. *To control rodents by:*

(a) Rodent-proofing the building. All openings to outside should be protected against the entrance of rats and mice; to do this properly,

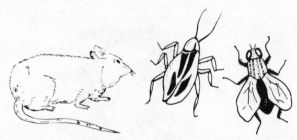

Figure 17. Keep them out!

the services of a professional are recommended. Rodents may squeeze through very small openings. A crack ½" wide will admit an adult mouse or a young rat. Doors should be self-closing.

(b) Exterminating rodents, by using the services of a professional exterminator.

(c) Keeping constant watch for signs of rodents' presence and activities. Signs of mice are their droppings (¼" in size) and gnawings on wood and food containers (sacks, crates, cartons); signs of rats are their fairly large pellets (½ to ¾"); their runways (characterized by a smudgy trail); and gnawings.

(d) Maintaining high standards of cleanliness at all times.

(e) Inspecting the premises every night for left-out food, open screens, open doors, and other entrances to the outside.

(f) Inspecting groceries, upon their delivery, for the presence of rodents.

The responsibilities of employees are to:

1. Not leave screen doors and other entrances to the outside open.
2. Replace the covers of garbage cans immediately after emptying the garbage, and close the door to the garbage room when leaving.
3. Clean up spills in kitchen, pantry, dining area, and storerooms immediately and thoroughly.
4. Never leave food unprotected.
5. Never leave food out overnight.
6. Report to supervisor the presence of vermin or signs of the activity of vermin.

SEPARATE STORAGE AREA FOR CLEANING MATERIALS, UTENSILS, AND EQUIPMENT

Cleaners such as detergents, soaps, ammonia, polishes, and other materials must be stored separately, never in the kitchen except that this requirement does not prohibit the convenient availability of detergents or sanitizers at utensil or dishwashing stations. Cleaning equipment and utensils (such as vacuum cleaners, brooms, buckets, mop trucks, and the like) must be stored separately.

The responsibilities of management are to:

1. Label containers of poisonous or toxic material prominently and distinctly according to the law for easy identification of contents.
2. Provide a separate, well-ventilated and well-lighted janitor area for storing cleaning materials, utensils and equipment.
3. Provide, in the janitor area, a separate sink for drawing water; making up cleaning solutions; washing out mops, sponges and other cleaning utensils; and emptying buckets.
4. Provide racks for drying cleaning utensils.
5. Schedule the area for cleaning.
6. Inspect the area for orderliness and cleanliness.
7. Inspect the area for improper storage of poisonous or toxic materials.

The responsibilities of employees are to:

1. Report to supervisor when supplies and utensils need replenishing and when equipment requires repair.
2. Keep janitor area orderly and clean.

REGULAR AND ADEQUATE CLEANING OF FACILITY

Cleaning of the foodservice facility is a fundamental and important part of the operation. Routine checks are made by inspectors from various county, state, and federal health departments to insure that high standards in sanitation are maintained.

THE PURPOSE OF CLEANING IS TO REMOVE SOIL

Soil is "matter out of place."
Soil or dirt can be dust, grime, grease, food particles, and even bacteria.
Types of soil fall in two basic categories:

(a) Soils that water will dissolve; these are the water-soluble soils. Examples are sugar, jelly, fruit juices, and salt.
(b) Soils that water will not dissolve; these are the water-insoluble soils and our most common forms of dirt. Examples are grease, oil, dust, food particles, and other solid matter, usually in various combinations.

WHAT IS A CLEAN SURFACE?

A surface is clean when it is free of soil; it is *bacteriologically clean when the majority of bacteria have been removed by sanitizing* the surface, using hot water or special chemical sanitizers. Therefore, when bacteriological cleanliness is necessary from a public-health point of view, sanitizing must be part of the cleaning operation. Cleaning and sanitizing may be performed in one operation, using special cleaning-sanitizing agents, if the soil is light. However, when soil is heavy, then the surface should be washed and rinsed prior to sanitizing. Heavy soil tends to deactivate the sanitizer.

HOW TO CLEAN

The responsibilities of management are to:

1. Set the standards. The local health department will advise a foodservice operator what sanitation standards are expected of the establishment.
2. Prepare a cleaning schedule.
3. Train personnel how to clean.
4. Provide adequate cleaning equipment, utensils, and materials.
5. Provide supervision to see that jobs are done properly.

The personnel must be informed regarding:

- What has to be cleaned.
- When a soiled item should be cleaned.
- How it should be cleaned.
- How to keep it clean until next use.

CLEANING SCHEDULE

It would be desirable if one could set up a cleaning schedule applicable to every foodservice establishment. This is impossible, since very few foodservice operations are alike. Some cleaning is daily routine and should be included in the daily work schedule of the employees. Some cleaning is weekly and should be scheduled on the day when the most employees are available or when the work load is lightest. Special cleaning, which is necessary a few times a year, may require specific scheduling of additional specialized help.

In general, *there should be no large-scale sweeping and cleaning while food is being prepared.* Dry sweeping raises dust, which contaminates the food. Therefore, only dustless methods of cleaning floors and walls should be used, such as vacuum cleaning, wet cleaning, or using dust-arresting sweeping compounds with brooms. Wet sweeping is apt to cause slipperiness and invites accidents, so caution must be used with this method. *However, spilled food should be removed from floors immediately,* dry spills by sweeping and wet spills by wiping.

Below is a list that indicates when specific areas should be cleaned. *This represents a minimum cleaning schedule.* Under certain circumstances more frequent cleaning may be necessary.

Daily

1. Floors, at least those of areas where traffic is heavy, as kitchens, bakeshops, pantries, dishwashing rooms, much-used walk-in refrigerators and vestibules, corridors, dining rooms, washrooms, toilet rooms.
2. Washbowls in kitchens, washrooms, and locker rooms.
3. Hood filters (if fryers and broilers are used daily)

Weekly

1. Floors in areas where traffic is only light. Much-used floors need to be cleaned more frequently, according to use
2. Hood

3. Walk-in refrigerators
4. Platforms
5. Duckboards
6. Garbage rooms
7. Windowsills, transoms

Monthly

1. Inside windows
2. Storerooms

Special

1. Outside windows
2. Walls, unless spattering requires frequent, regular cleaning
3. Light fixtures
4. Venetian blinds
5. Freezers

DIRECTIONS

Cleaning procedures must be written out as specific directions. Management or other key personnel are responsible for providing these, and employees should be instructed in (1) *the use of cleaning equipment and utensils* and (2) *the use of cleaners.* Stressed should be the importance of using the prescribed cleaners in the prescribed strength for the specific jobs. It is most important to READ LABELS AND FOLLOW MANUFACTURER'S DIRECTIONS.

Measuring devices should be provided for preparing the solutions exactly as prescribed; employees should also be instructed in (3) *the care of cleaning equipment and utensils after use,* such as the cleaning and drying of buckets, mops, brushes, sponges, cloths, and the like. Cleaning equipment, utensils, and materials must never be stored with food. A special area equipped with water, a sink, and drying racks should be provided where equipment and utensils can be cared for and stored.

FLOOR CARE

Floors require different care according to the material of which they are made.

Some general rules are as follows:

1. Wipe up spills promptly before they are tracked over the floor.
2. Determine the desirable frequency of cleaning; frequency of general cleaning depends on factors such as the extent of traffic and the degree and type of soil. Floors subjected to heavy traffic, food spills, and the like, must be cleaned daily.
3. Hard floors (except wood floors) take flooding with water; examples of hard floors are marble, terrazzo, ceramic tile, and cement floors. Wood floors* made of hard wood are hard floors, but they do not tolerate flooding with water or water-base cleaning solution.
4. Soft floors do not take flooding with water; examples of soft floors are asphalt, linoleum, rubber tile, and cork.
5. Floors need to be cleaned of gross litter and sticky and caked material and swept before they are wet-cleaned. Use a putty knife to remove caked and sticky soil.

Suggestions for the type of cleaner to be used for different floors are given in Table 2.

BROOM-SWEEPING

Utensils and materials needed:

Long-handled broom
Dustpans and hand brush

---

*Wood is no longer used for kitchen floors, but hard woods (oak, birch, and maple) are still being used in dining rooms.

Table 2. Floor cleaning

Type of Flooring	Hardness	Where Used	Advantages	Disadvantages	Cleaner
Terrazzo	Hard marble chips in cement	Kitchen; dining rooms	Durable	Damaged by alkalies; stained by grease unless well sealed	Use mild cleaners; scrub; rinse; mop dry
Ceramic tile	Hard		Resists grease stains if glazed; durable	Unglazed tile easily stained by grease; spaces between tile difficult to clean; slippery when wet	Alkaline cleaners are safe; scrub to clean between tiles; rinse; mop dry
Concrete	Hard		Durable	Gives off dust; becomes stained by grease unless well sealed	Alkaline cleaners are safe; scrub, rinse (sealing concrete will make it more cleanable)
Asphalt tile	Soft	Dining rooms	Durable; non-slippery even when wet	Poor resistance to grease	Mildly alkaline to neutral cleaners; mop; rinse; mop dry
Rubber tile	Soft		Durable	Damaged by grease	Mild cleaners; mop; rinse; mop dry; NEVER USE STRONG CLEANERS
Vinyl	Soft		Nonslippery even when wet; durable; resistant to grease		Mildly alkaline to neutral cleaner; mop; rinse; mop dry
Vinyl asbestos	Soft		Durable; resistant to grease		Alkaline cleaners safe; mop; rinse; mop dry
Cork tile	Soft			Easily damaged by sharp objects; grease will penetrate unless well sealed	Use special cleaners for cork floors (sealing cork tile makes floors more cleanable)

Putty knife
Dust-arresting sweeping compound
Coarse (nail) comb

Procedure

1. Pick up gross litter; pry up sticky material with a putty knife.
2. Spread area to be cleaned with dust-arresting sweeping compound.
3. Start sweeping where the area is entered.
4. Sweep under and around large pieces of equipment; use a hand brush for areas that cannot be reached with a broom.
5. Do not accumulate dirt in large piles; sweep it into small piles to be picked up later with a counter brush and dustpan.
6. Clean brushes with a coarse (nail) comb. If they are soiled, wash them in water and detergent, rinse, and hang up to dry. Clean the dustpan and putty knife.

DUST-MOPPING

Dust-mopping raises less dust than broom-sweeping. Dust-mopping is more efficient than broom-sweeping on such floors as wood, asphalt tile, vinyl tile, linoleum, terazzo, and on floors that have been sealed. Dust mops should be pretreated with dust-arresting compound before use.

Utensils and materials needed:

Pretreated dust mop
Dustpan and hand brush
Putty knife

Procedure

1. Pick up gross litter; pry up sticky material with a putty knife.
2. Move easily movable equipment and furniture out of the way for easier cleaning.
3. Mop as closely as possible around nonmovable equipment, making sure that strokes overlap and all areas get dusted.
4. Frequently shake out the mop into small piles to be picked up later; use a dustpan and handbrush to pick up piles.
5. Clean dust mop with a stiff brush; if quite soiled, have it laundered and pretreated. Hang mop up to air.
6. Dust mopping is not recommended for kitchens; may be used in dining rooms and reception areas.

WET-MOPPING

Before wet-mopping, broom-sweep or dust-mop first. Move easily movable equipment and furniture out of the way.

Utensils and materials needed:

Broom or pretreated dust mop
Dustpan and hand brush
Putty knife
Two buckets, one with a wringer
Two mops, one for cleaning and one for rinsing and drying
Cleaning ingredient (should not damage floor)

Procedure

1. Prepare the cleaning solution. FOLLOW MANUFACTURER'S INSTRUCTIONS EXACTLY.
2. Broom-sweep or dust-mop the area first (see instructions above).
3. Start mopping at the rear of floor to be washed, working backward and parallel to the wall. Clean a small area (approximately 10 × 10 feet) at one time. Mop corners by hand or use the heel of the mop.
4. Rinse the cleaned section with a rinse mop; rinse the mop, wring it out, and use it to dry the cleaned and rinsed area.
5. Repeat the wash-rinse-dry operation, until the entire floor is clean. *Caution:* Do not allow rinse water to get dirty; change it frequently.
6. Clean the dust mop or broom, buckets, handbrush, dustpan, and putty knife. Clean the mop in heavy-duty detergent or cleaner-sanitizer, rinse, and hang it up to dry.

SCRUBBING (HAND OR MACHINE)

Before scrubbing, dust-mop the floor first. Move easily movable equipment out of the way. (See directions for dust-mopping.)

Equipment, utensils, and materials needed:

Dust mop, dustpan, hand brush, putty knife
Two wet mops, one for applying cleaning solution and one for rinsing and drying
Scrub brushes for scrubbing by hand, or automatic scrubbing equipment
Two buckets, one for the cleaning solution and one for rinse water

Procedure

1. Prepare the cleaning solution. FOLLOW MANUFACTURER'S INSTRUCTIONS EXACTLY.
2. Dust-mop the area first (see instructions above).
3. Wet-mop small sections (approximately 10′ × 10′) at one time.
4. Scrub by hand or machine.
5. Rinse each area twice; dry.
6. Always scrub corners and the strip along the baseboard and the baseboards by hand; clean up spatters on baseboards. *Caution:* Change wash and rinse water frequently.
7. Continue cleaner application-scrub-double rinse-dry sequence until the floor is done.
8 Clean mechanical equipment following the directions of the manufacturer. *Caution:* Never rest the scrubbing machine on the brushes, since it flattens and ruins them. Clean the mops, handbrushes, dustpan, and putty knife. Clean the buckets. Hang the wet mop and brushes to dry.

CARE OF WALLS

All walls must be dusted as part of routine cleaning. Walls that become heavily soiled must also be washed. The frequency of washing walls must be judged by those responsible for scheduling cleaning. Walls that are spattered daily must be cleaned daily; walls that stay clean for weeks need occasional cleaning. No hard-and-fast rule is possible except that *walls must be kept clean and free of dust, moisture, grease, grime, and mildew.* Mildew can be a real problem in areas where moisture is high.

To be cleanable, walls and ceilings should be smooth and without cracks and crevices. Walls and ceilings should be vermin proof. Light-color walls make dirt more visible. Painted walls should be repainted regularly. Papered walls should be kept in a good state of upkeep, since loose paper offers hiding places for vermin.

How to Wash Walls and Woodwork

1. Dust painted walls before washing them, using, preferably, a vacuum cleaner or pretreated dust mop; when dusting by broom or brush, dust is apt to be spread about. If broom or brush must be used, cover with a clean cloth. To remove cobwebs, use a lifting motion, because they are apt to be sticky and could cause streaks on wall if brushed downward.
2. Wash walls and woodwork with a cleaner that will not damage the surface.

Tile Walls

Utensils and materials needed:

Two buckets, one for washing and one for rinsing
Soft scrub brush
Sponges, cloths, or sponge mop with squeegee
Nonabrasive* cleansing powder (should not damage the tile)
Neutral detergent or other cleaner

Procedure

1. Wash the tiles with a sponge or cloth dipped in a warm, mild cleaning solution. Use a soft brush to clean between the tiles. If the walls have become grimy, a *non*abrasive cleansing powder may be used. However, frequent use of cleansing powder is damaging to tiles. Work from the bottom up to prevent streaking.
2. Clean a small area at one time, then rinse.
3. Rinse with clean warm water; change the rinse water frequently.
4. Dry with a soft, dry cloth.
5. Repeat the sequence wash-rinse dry until the entire wall is clean.

Painted Walls and Woodwork

General warning:

1. Painted surfaces are much more delicate than tile. *Do not use abrasives. Use a mild detergent or cleaner,* in proper dilution. READ THE LABEL ON THE PRODUCT. FOLLOW MANUFACTURER'S DIRECTIONS.
2. Rinse after using the cleaner, even if the directions on the label of the cleaner should state that rinsing may be omitted. Rinsing reduces the possibility of damage to paint.

Utensils and supplies needed:

Two buckets
Sponges, cloths, or sponge mop with squeegee
Warm solution of neutral detergent or mild soap
Warm water for rinsing sponges and cloths

---

*Nonabrasive means nonscratchy.

Procedure

1. Wash painted surfaces with a sponge dipped in cleaning solution. Work from the bottom up to avoid streaking. Streaks may form when the cleaning solution runs down on the soiled portion and dries there. Use cleaning solution on a small area at a time, then rinse and dry. Repeat the process until the entire surface is done.
2. Rinse. Wipe with a damp sponge or a cloth wrung out in clean warm water. *Caution:* Change the rinse water frequently.
3. Dry with a soft dry cloth.

Wallpaper (Washable)

1. Test the paper to ascertain whether the cleaner avoids roughing the surface, making the colors blend, or loosening the paper from the wall.
2. Follow the procedure for painted walls, but *never rub dry*; rather, gently pat dry. *Caution:* Great care is needed to avoid ruining the surface. Work very gently. Use as little water as possible. DO NOT RUB. The method described for nonwashable paper can be used for sensitive washable paper.

Wallpaper (Nonwashable)

1. Use wallpaper cleaner that looks like a lump of dough.
2. FOLLOW DIRECTIONS EXACTLY. The product is stroked over the paper, and dust particles and greasy soil should adhere to it. *Caution:* Knead the "dough" frequently to bring out clean surfaces; do not allow old soil to be spread back onto the paper.

HOOD AND HOOD FILTER

Hoods over stoves and fryers collect moist, greasy air. Hoods may become a sanitary hazard if they are not kept clean, because the drip from the hood may fall onto food and contaminate it. Hood filters do not function properly if they are gummed up with grease. Greasy hoods are also liable to cause fires, and are therefore dangerous from a safety point of view.

Filter.   Wash the hood filter in the automatic dishwasher. Do this at least once a week, and daily if fryers and broilers are used daily.

Hood.   Wash the hood, using the procedure described for walls. Use a detergent solution to remove grime. Rinse and dry. *Warning:* Be sure to use plenty of "elbow grease" to remove sticky accumulations from the channels each time the hood is washed. Hoods should be washed at least once a week and professionally cleaned once a year and treated with a fire retardant.

HOW TO CLEAN WASHBOWLS, TOILET BOWLS, AND URINALS

Washbowls. The surface of washbowls may be scratched easily if abrasives are used. If soap scum has built up, a nonabrasive cleansing powder or a water softener such as Calgon may be used.

For regular care, washbowls should be cleaned simply with a solution of detergent and warm water.

Utensils and materials needed:

Detergent or other mild cleaner
Sponge
Cloth

Procedure

1. Rinse the basin by running the water for several seconds.
2. Wash out the basin with a sponge and detergent solution; remove built-up scum or stains with a nonabrasive cleaner or Calgon solution.
3. Rinse with clear water.
4. Dry with a soft cloth.
5. Wipe the chrome fixtures clean.

Toilet Bowls and Urinals. Toilet bowls and urinals are more resistant to chemical cleaners than are washbowls; therefore, certain chemicals can safely be used to clean them. *Use cleaners especially designed for toilets and urinals.* Toilet bowls and urinals should also be sanitized. FOLLOW DIRECTIONS EXACTLY. *Caution:* Never use toilet bowl and urinal cleaners on washbowls, since they may permanently damage the washbowls and render them difficult to keep sanitary.

Utensils and materials needed:

Special cleaner and sanitizer (detergent, disinfectant concentrate)
Special brushes
Cloths

Procedure

1. Flush commodes or urinals.
2. Pour the bowl cleaner into the bowls; if the bowls are stained, allow the cleaner to act for a while.
3. Using a bowl brush or mop, clean the entire bowl thoroughly and flush.
4. Using special applicators, apply disinfectant-cleaner solution to the bowl or urinal, inside and out.

5. Using a special applicator, wipe the tops and bottoms of toilet seats with a disinfectant-cleaner solution and wipe them dry with a cloth.

WINDOWS, MIRRORS AND LIGHTS

Frequent cleaning prevents build-up of greasy film; set up cleaning schedule to prevent this from happening.

Equipment and materials needed:

For daily routine vacuum-dusting or cleaning with water and a damp cloth, paper toweling or chamois is satisfactory.

Procedure

1. Use a sponge, chamois, or soft cloth; moisten it with water or cleaning solution; wring dry.
2. Wash the soiled surfaces.
3. Wipe with a dry, lint-free cloth, or a damp sponge, or a squeegee. When using a squeegee, stroke from the top to the bottom and wipe the squeegee after each stroke.

DO NOT WASH LIGHT FIXTURES WHILE THEY ARE STILL CONNECTED TO THE ELECTRIC OUTLET. Remove tubes and bulbs and clean with the window-washing solution.

SECTION B

Sanitary Equipment and Utensils

As stated by Fischback, a sanitation consultant: "a clean wholesome meal is guaranteed the customer when you offer the menu. This is the responsibility of a foodservice operator. The food must be palatable, prepared by clean people, in clean utensils, with clean equipment. It should be served on clean dishes with clean tableware from a clean table on a clean floor, in clean surroundings."

A foodservice operator can no longer fulfill the responsibility to the customer by providing quality food and service. Quality sanitation must also be provided. Therefore foodservice managers and employees are really custodians of the health and welfare of their customers.

The primary objective of a cleaning operation in a public eating establishment is the control of bacteria. This is necessary (1) to prevent food spoilage, (2) to reduce health hazards, and (3) to control odors.

All food-preparation equipment and utensils may be contaminated with bacteria capable of producing disease and, therefore, this equipment constitutes a potential public-health hazard. Thorough cleaning and proper sanitizing are essential. Cleaning involves removing the soil, but sanitizing involves reducing the bacterial count to a safe level.

Fundamentally, the main cleaning agent is water. The use of cleaning compounds and friction helps water to do a better job. Cleaning is a practical application of chemistry. Since chemistry is one of the few exact sciences, successful cleaning involves the use of the proper amounts of detergent in clean water at the proper temperatures [110 to 120 F (43.4 to 49 C) for handwashing and 130 to 140 F (54.4 to 60 C) for machine washing].

The purpose of the detergent is to loosen the soil. Friction must be used to remove it. Friction is accomplished by cleaning tools (brushes and elbow grease) or cutting action by water sprays. Then the soil must be rinsed off or it will dry back onto the equipment surface. A clean surface is necessary for sanitizing because soil can inactivate the sanitizer. Clean water, at 170 F

(76.6 C) for 1 minute or at 180 F (82 C) for 15 seconds, is necessary to rinse away soil and sanitize the equipment. If water at these temperatures is not available for the final rinse, then immersion for at least one minute in a clean solution containing at least 50 parts per million of available chlorine as a hypochlorite and at a temperature of at least 75 F (24 C) is required. Finally, any object that has been cleaned and sanitized must be placed on a clean surface and allowed to air dry.

A list of other approved chemical sanitizers may be found in the Food Service Sanitation Manual (1976). When chemicals are used for sanitization, they should be in the proper concentration and a test kit or other device that accurately measures the parts per million concentration of the solution should be provided and used.

Before cleaning a specific type of equipment (utensil or surface) consideration must be given to the following factors.

1. *The nature of the water being used.* The water should be properly conditioned to make the detergent more effective and to eliminate the deposit of any minerals present in the water.
2. *The type of soil to be removed.* Soil may be grease, protein, mineral, or carbon. Thus, different types of cleaning compounds or different water temperatures may be required.
3. *The corrosion resistance of the material being cleaned.* This will determine the type and amount of friction that can be applied.
4. *The type of cleaner being employed.* Soap may leave a greasy film because of poor rinsing qualities; therefore, it is not recommended for general use. An abrasive may roughen the surface and increase the possibility of future soil. Most detergents, on the other hand, will do a good job if they are properly formulated and applied.
5. *The condition of the soil.* Soil may be fresh, soft, dried, or baked-on. Temperature plays an important role because soil that dries on at a low temperature is usually easy to remove, while baked-on residues create a highly specialized cleaning problem.

Proper sanitation of equipment begins at the time of purchase. The equipment should be designed for easy cleaning. Movable parts (particularly those that come into contact with food) should be removed easily without the need of special tools. The remaining shell should be made of smooth, noncorrosive, nonchip material, free from screw heads, square corners, or cracks in which soil, grease, or food particles can lodge. It is as important to check the cleanability and the time required for cleaning a piece of equipment before purchase as it is to check the quality and capacity of production. If the equipment carries the National Sanitation Seal of

Approval, the manager can be sure that it has met the requirements for cleanability; however, the time required for cleaning must be determined by test, preferably under actual operating conditions.

Proper installation is another major requirement for equipment sanitation. Stationary equipment should be installed far enough from the wall so the worker can clean behind it. Mobile models of most kitchen equipment are now available and should be considered for any new installations. Special attention must be given to water and utility line connections. A labyrinth of pipes or electrical cords creates a difficult cleaning problem and may constitute a safety hazard for the cleaner.

Good sanitation techniques in a food operation do not just happen. They result from a well-planned program that requires (1) the basic knowledge of what is involved in safe food handling, (2) the ability to translate this knowledge into practical and workable procedures to be followed in the daily routine of serving food that will ensure a wholesome product, and (3) a constant effort on the part of management not only to maintain the standards that have been set but also to emphasize their importance.

Foodservice equipment can be classified under four major categories, listed below, and each category has its own sanitation significance.

1. *Equipment that comes into actual physical contact with the food material being processed.* This category includes all cutting, dicing, slicing, grinding, and mixing equipment, and all utensils, dishware, and chopping boards. Because of the physical contact between food and equipment, the possibility of microbial contamination is high; therefore, it is extremely important to use equipment that can be dismantled easily so contact parts can be taken to the scullery or the dishwasher for proper cleaning and sanitizing.

2. *Primary equipment used in cooking or holding food.* Ovens, ranges, broilers, and fryers are examples. Here the food does not come into actual contact with the equipment or, if it does, the contact is at temperatures that make the problem of microbial contamination negligible. The sanitation concern becomes one of odor control, esthetics, and equipment efficiency.

Soiled equipment is always a potential source of undesirable odor; where heat is involved, these odors are accentuated. The acrid odor emanating from a soiled deep-fat fryer has earned many a foodservice operation the title of "The Greasy Spoon." Soiled equipment rarely works at optimum efficiency. A layer of burnt-on food on the oven hearth acts as an insulator and reduces oven temperature. It is poor economy to buy an oven with good thermostatic controls and then allow soil to destroy its efficiency. Proper maintenance of equipment not only controls odor but also cuts down on repair bills and extends the life of the equipment.

3. *Equipment used in cleaning other equipment.* Dishwashers and pot-and-pan sinks are common examples of this category of equipment. The importance of cleanliness should be obvious but, all too frequently, the proper cleaning of these machines is neglected because of an assumption that the large amounts of water and detergent used in the normal course of operation make these machines self-cleaning. Remember that dishes or utensils cannot be any cleaner than the machine in which they are washed. Mineral deposits (which result from improperly conditioned water) and scrap trays filled with food residue provide excellent breeding spots for microorganisms. Scrap trays need to be emptied and washed after each meal, and mineral deposits should be removed regularly. Partially plugged wash-or-rinse arms in the machines destroy the uniform wash-and-rinse pattern necessary to produce clean dishes. They should be inspected daily and removed for cleaning when necessary.

Effective manual pot-and-pan washing requires a three-compartment sink with double drainboards. One drainboard provides a stacking surface for soiled pots and pans, and one is used for the air-drying of clean utensils after the sanitizing procedure. A double-walled compartment with a drain that has its overflow rim at the normal water level of the sink (between the wash and rinse compartments) permits the "skimming" of sude or grease films from the surface of the rinse water. Since grease and suds float, the skimming device allows them to be flushed off. Thus the rinse remains free of suds and grease film throughout the cleaning process.

4. *Transportation and mobile storage equipment.* This category of equipment, like primary equipment, does not usually come in contact with food, but does come in contact with pans, utensils, and dishes that, in turn, come in contact with food. Mobile dish-storage equipment, such as lowerators and dish carts, need special attention. Dishes are only as clean as the equipment in which they are stored.

Management must keep in mind that equipment designed for easy cleaning will be clean only to the degree of cleanliness demanded by management and exercised by the worker. The worker, in addition to being provided the necessary tools and cleaning compounds, must be trained to do the job properly. A demonstration by the manager or the supervisor of how the job should be done, followed by a practice period under supervision, will assure management that the worker has learned to do the job correctly. A written cleaning procedure is a useful training tool and serves as a standard for checking the quality of performance. Daily inspection by management of all routine cleaning is extremely important to the success of the program.

A checklist of all items that are to be cleaned should be developed. Since it is safe to assume that all equipment in the kitchen needs to be cleaned at some time, a systematic method of listing the equipment should be used. The operator should start at one end of the kitchen and move to the other end, listing all equipment and grouping like equipment together. This will eliminate the possibility of overlooking some items. The second step is to determine how often each item on the list should be cleaned. The third step is to develop a cleaning schedule in such a manner that all items on the checklist are delegated to a definite person at a definite time. These cleaning assignments should then be incorporated into the employee work schedules. Written cleaning schedules and cleaning procedures are the tools that management uses to control the sanitation program. The cleaning schedule determines who is responsible for the job, and the cleaning procedure prescribes how the work is to be accomplished. Daily follow-up is a "must" for an effective sanitation program. The format of all cleaning procedures should:

1. Give the name of the equipment being cleaned.
2. Have three columns headed "when," "how," and "use."
3. Include all steps (written as briefly as possible) that are necessary to properly dismantle, clean, and reassemble the equipment.
4. Take into consideration the safety precautions necessary to make the equipment safe to clean.
5. Specify the amount of detergent to use and the temperature of water to use for washing, rinsing and sanitizing.
6. Stress the need for a clean surface for air drying and clean hands for reassembling equipment.
7. Provide a separate procedure if daily cleaning differs from weekly or bimonthly cleaning.

The following cleaning procedures are offered as examples of the type that can serve both as a training tool and a standard for evaluating the effectiveness of performance. These procedures are based on a specific model of each type of equipment. Remember that there are many different models available. Therfore, the dismantling and reassembling instructions may need to be changed to conform to the specific model in use in any given food service.

The proper proportion of detergent to water required for each cleaning job should be specified. Because of the variety of detergents available, it has not been possible to specify the amounts of detergents to be used in these generalized procedures. Each operator must determine the appropriate amount of detergent available in the operation for each cleaning job. Employees should be provided with convenient measuring and weighing equipment so the correct amount of detergent will be used. If more than one type of cleaning material is available, the name of the correct cleaning material and the amount to be used should be part of the cleaning procedure.

Procedure 1. How to Clean Food Slicer

When	How	Use
After each use	1. Turn off machine	
	2. Remove electric cord from socket	
	3. Set blade control to zero	
	4. Remove meat carriage	
	(a) Turn knob at bottom of carriage	
	5. Remove the back blade guard	
	(a) Loosen knob on the guard	
	6. Remove the top blade guard	
	(a) Loosen knob at center of blade	
	7. Take parts to pot-and-pan sink, scrub	Hot machine detergent solution, gong brush
	8. Rinse	Clean hot water, 170 F (76.6 C) for 1 minute. Use double S hook to remove parts from hot water
	9. Allow parts to air dry on clean surface	
	10. Wash blade and machine shell	Use damp bunched cloth[a] dipped in hot machine detergent solution
	Caution: PROCEED WITH CARE WHILE BLADE IS EXPOSED	
	11. Rinse	Clean hot water, clean bunched cloth
	12. Sanitize blade and machine shell, allow to air dry	Clean water, chemical sanitizer, clean bunched cloth
	13. Replace front blade guard immediately after cleaning shell	
	(a) Tighten knob	
	14. Replace back blade guard	
	(a) Tighten knob	

Procedure 1. (Continued)

When	How	Use
	15. Replace meat carriage (a) Tighten knob 16. Leave blade control at zero 17. Replace electric cord into socket	

[a] *Fold cloth to several thicknesses.*

Procedure 2. How to Clean Vegetable Chopper

When	How	Use

After each use	1. Turn off switch, wait until blades have stopped
	2. Release safety lock [a]
	3. Turn cover lock one fourth turn and pull up
	4. Raise cover to upright position
	5. Remove blade guide
	    (a) Lift up
	    (b) Place in bowl
	6. Remove blade
	    (a) Remove knob on blade shaft
	    (b) Grasp blade at the dull surfaces near the shaft
	    (c) Pull blade off shaft
	    (d) Place blade into bowl
	Caution: HANDLE BLADE WITH CARE
	7. Lift off cover
	8. Remove bowl
	    (a) Give bowl one half turn clockwise and lift	

	9. Take cover, bowl, blade guide, and blade to sink and scrub *Caution:* NEVER PLACE BLADE IN SINK; LEAVE ON DRAINBOARD UNTIL READY FOR WASHING AND RINSING	Hot machine detergent solution, gong brush
	10. Rinse parts, allow to air dry on clean surface	Clean hot water, 170 F (76.6 C) for 1 minute

(continued)

Procedure 2. (Continued)

When	How	Use
	11. Scrub shell of machine	Clean damp cloth dipped into hot machine detergent solution
	12. Rinse shell, wipe dry	Clean hot water, chemical sanitizer, clean dry cloth
	13. Reassemble chopper (a) Replace bowl, place on spindle, give half turn counterclockwise (b) Replace cover, leave in upright position (c) Replace blade guide and blade (d) Close cover, put on safety lock	

[a] *In new models, safety lock is part of switch.*

Procedure 3. How to Clean Vertical Cutter/Mixer (Figure 18)

When	How	Use
After each use	1. Turn cutter motor to "off" position (a) Open inspection cover on top (b) Wait until blades have stopped 2. Release cover lock 3. Raise cover 4. Fill bowl half full of hot machine detergent solution	Hot water, detergent
	5. Close and lock cover (a) Close inspection cover 6. Turn cutter motor to "on" position Allow to run 30 seconds 7. Turn cutter motor to "off" position (a) Open inspection cover (b) Wait until blades have stopped 8. Release cover lock	

Procedure 3. (Continued)

When	How	Use
	9. Raise cover	
	10. Tilt bowl to remove water	Large pail or floor drain
	(a) Remove locking pin at base of bowl	
	(b) Release lever to tilt kettle	
	(c) Lock lever to hold bowl in new position	
	11. Rinse inside and outside of bowl, cover, and frame	Hose, hot water
	12. Wipe outside of machine, frame, and cover until dry; allow inside to air dry	Clean dry cloth
	13. Return bowl to upright position	
	(a) Remove locking pin	
	(b) Release lever to raise bowl	
	(c) Lock lever to help hold bowl	
	14. Close cover, do not lock	
	(a) Leave inspection cover open	
Daily	1. Turn cutter motor to "off" position	
	(a) Open inspection cover on top	
	(b) Wait until blades have stopped	
	2. Remove mixing baffle	
	(a) Loosen nut on top of cover and withdraw handle	
	(b) Release cover lock	
	(c) Raise cover to upright position	
	(d) Withdraw mixing baffle and shaft	
	3. Remove plastic inspection cover and gasket	
	4. Remove blade shaft	
	(a) Turn blade shaft lock clockwise to remove	
	(b) Pull up on blade shaft	

(continued)

Figure 18. Vertical vegetable cutter. (Courtesy of Hobart Mfg. Co.)

Procedure 3. (Continued)

When	How	Use
	5. Take parts to sink and scrub *Caution:* LEAVE BLADESHAFT ON DRAINBOARD UNTIL READY FOR WASHING AND RINSING	Hot machine detergent solution, gong brush
	6. Rinse parts and allow to air dry on clean surface	Clean hot water, 170 F (76.6 C) for 1 minute
	7. Fill bowl half full of hot machine detergent solution	Hot water and detergent
	8. Scrub bowl, motor stem, and inside and outside of cover; scrub side of bowl and frame	Gong brush
	9. Tilt bowl to remove water (a) Remove locking pin at base of bowl (b) Release lever to tilt kettle (c) Tighten lever with bowl in new position (d) Replace locking pin to hold position	Large pail or floor drain
	10. Rinse inside and outside of bowl, cover, and frame; wipe outside, allow inside to air dry	Hose, hot water, clean dry cloth
	11. Return bowl to upright position (a) Remove locking pin at base of bowl (b) Release lever to raise bowl (c) Lock lever to help hold bowl (d) Replace locking pin	
	12. Reassemble parts (a) Replace blade shaft, lock in position (screw on tight) (b) Replace gasket (c) Replace mixing baffle and shaft into cover (d) Close cover (e) Replace plastic inspection cover (f) Replace handle (g) Replace nut on top of cover, tighten	
	13. Leave cover closed, do not lock (a) Partially open inspection cover	

Procedure 4. How to Clean Food Mixer

When	How	Use
After each use	1. Turn off machine	
	2. Remove bowl and beaters, take to pot-and-pan sink for cleaning	Hot machine detergent solution, short handled gong brush
	3. Rinse, keep immersed for 1 minute, air dry	Clean hot running water 170–180 F (76.6–82 C)
Daily	4. Scrub machine (a) Beater shaft (b) Bowl saddle (c) Shell and base	Hot machine detergent solution, gong brush, clean cloth
	5. Rinse	Clean damp cloth wrung out in clean hot water with chemical sanitizer added.
	6. Dry	Clean dry cloth
	7. Oil as per directions if required	Machine oil

Procedure 5. How to Clean Steak Machine

When	How	Use
After each use	1. Turn off machine	
	2. Remove electric cord from socket	
	3. Remove guard	
	4. Remove cutter roll assembly (a) Loosen locks (b) Pull entire roll assembly to right to remove from machine	
	5. Take guard and cutter roll assembly to sink (a) Remove handle (b) Remove blade guides, handle cutter roll with care	Pot-and-pan sink or sink in meat shop
	(c) Clean excess meat particles from cutter roll	Special cleaning fork
	Caution: PLACE CUTTER ROLL ON DRAINBOARD UNTIL READY FOR WASHING	

Procedure 5. (Continued)

When	How	Use
	6. Scrub all parts	Hot machine detergent solution, stiff brush
	7. Put all parts into wire basket and immerse in rinse	Clean, hot water, 170 F (76.6 C) for 1 minute
	8. Allow to air dry in basket on clean surface	
	9. Clean shell of machine	Clean damp cloth, dipped in hot detergent solution
	10. Rinse	Clean hot water chemical sanitizer, cloth
	11. Dry shell	Clean dry cloth
	12. Reassemble cutter roll: (a) Replace blade guides (b) Replace handles (c) Replace roll assembly into socket (d) Lock roll assembly (e) Replace guard	
	13. Replace electric cord into socket	

Procedure 6. How to Clean Power Meat Saw

When	How	Use
After each use	1. Turn off power 2. Remove parts: (a) *Pan.* Open bottom door to remove pan (b) *Small table section.* Tilt and slide guage plate to the back. Lift out small table section (c) *Blade.* Turn blade control clockwise to loosen tension. Pull back guard down to swing free of blade. Coil blade (d) *Meat scrapers.* Release sprng clip and slide from shaft	

(continued)

Procedure 6. (Continued)

When	How	Use
	(e) *Blade wheels (upper and lower).* Loosen knob and remove	
	(f) *Moving table.* Release safety catch below table and slide off	
	(g) *Lower blade roller.* Release thumb screw in front at base and remove	
	(h) *Blade guide.* Remove setscrew on top and pull down to release; replace setscrew while washing blade guide	
	3. Take all parts to sink for washing	Hot machine detergent solution, gong brush and small vegetable brush
	4. Rinse:	Clean hot water, 170 F (76.6 C)
	(a) Place small parts in wire basket and immerse	Long-handled wire basket
	(b) Immerse large parts	
	(c) Leave all parts immersed at least one minute	
	(d) Remove parts to clean surface; allow to air dry	Wire basket and double S hook
	5. Clean machine shell	
	(a) Remove all meat particles	Hand
	(b) Scrub down machine	Hot machine detergent solution, clean cloth and brush
	6. Rinse	Clean hot water, chemical sanitizer, clean cloth
	7. Wipe dry	Clean dry cloth
	8. Replace parts in reverse order	

Procedure 7. How to Clean Food Grinder

When	How	Use
After each use	1. Turn off machine	
	2. Remove electric cord from socket	
	3. Loosen ring on grinder, release safety clamp	

Procedure 7. (Continued)

When	How	Use
	4. Remove pan and grinder head and take to pot-and-pan sink (a) Remove ring (b) Take out plate, remove food particles (c) Remove knife (d) Remove feed screw from cylinder	Ice pick and/or stiff brush, hand
	5. Wash parts in pot-and-pan sink, give special attention to threads and grooves in cylinder and ring	Hot machine detergent solution, gong brush, ice pick
	Caution: DO NOT PUT KNIFE INTO SINK; LEAVE ON DRAINBOARD UNTIL READY FOR WASHING	
	6. Put all clean parts in wire basket and immerse in rinse	Clean hot water, 170 F (76.6 C) for 1 minute
	7. Remove to clean dry surface and allow to air dry	
	8. Oil threads on ring and cylinder	Salad oil
	9. Clean outside of grinder shell	Hot machine detergent solution, clean cloth
	10. Rinse shell	Clean hot water, chemical sanitizer, clean cloth
	11. Wipe dry	Clean dry cloth
	12. Store grinder parts in appropriate place [a]	

[a] *Do not reassemble unless the next job requires same size plate.*

Procedure 8. How to Clean Bench Can Opener

When	How	Use
Daily	1. Remove shank to pot-and-pan sink	
	2. Scrub, give special attention to blade and moving parts	Hot machine detergent solution, gong brush
	3. Rinse	Clean hot water, 170 F (76.6 C) for 1 minute
	4. Allow to air dry on clean surface	

(continued)

Procedure 8. (Continued)

When	How	Use
	5. Inspect blade and replace when notched	
	6. Scrub base plate attached to table	Hot machine detergent solution, gong brush
	7. Rinse base plate	Clean hot water, chemical sanitizer, clean cloth, air dry
	8. Replace shank	
Periodically	1. Remove base plate from table	Screw driver
	(a) Scrape area free of all food particles	Putty knife
	(b) Scrub area	Hot machine detergent solution, gong brush
	(c) Rinse	Clean hot water, chemical sanitizer, clean cloth, air dry
	(d) Replace base plate	Screw driver

Procedure 9. How to Clean Platform Scales

When	How	Use
Daily	1. Remove weights	To cart or table
	2. Scrub platform	Hot machine detergent solution, gong brush
	3. Rinse	Clean hot water
	4. Wipe dry	Clean dry cloth
	5. Replace weights	
Weekly	1. Remove weights	To cart or tables
	2. Scrub platform, frame, weights, and wheels (tip scale to side to clean wheels)	Hot machine detergent solution, gong brush, clean cloth
	3. Rinse	Clean hot water, chemical sanitizer, clean cloth
	4. Wipe dry	Clean dry cloth
	5. Replace weights	

Procedure 10. How to Clean Table Scales (Even Balance)

When	How	Use
After each use	1. Remove scale pan	
	2. Brush all dry particles from scale frame and surrounding table top	Clean dry cloth
	3. Wash scales, scale pan, and table top beneath and surrounding scales	Hot machine detergent solution, gong brush, clean cloth
	4. Rinse and sanitize scale pan [a]	Clean hot water, chemical sanitizer
	5. Allow scale pan to air dry Wipe scale frame dry	Clean dry cloth

[a] *Scale pan can be washed in pot sink and sanitized with 170 F (76.6 C) water for 1 minute.*

Procedure 11. How to Clean Portion-Control Scales

When	How	Use
After each use	1. Brush off any food particles on platform	Clean damp cloth
	2. Scrub platform and frame; give special attention to notched balance beam	Hot machine detergent solution, gong brush, clean cloth
	3. Rinse and sanitize platform, allow platform to air dry	Clean hot water, chemical sanitizer, clean cloth
	4. Wipe scale frame dry	Clean dry cloth
	5. Set scale on clean surface	
	6. Scrub table top	Hot machine detergent solution, gong brush, clean cloth
	7. Rinse table top and wipe dry	Clean hot water, clean dry cloth
	8. Return scales to proper position	

Procedure 12. How to Clean Chopping Boards

When	How	Use
After each use	1. Take chopping board to pot and pan sink	
	2. Scrub immediately or place on soiled pan drainboard	Hot machine detergent solution, gong brush
	Caution: DO NOT SOAK CHOPPING BOARD	
	3. Rinse	Clean hot running water
	4. Sanitize	Clean hot water, chemical sanitizer
	5. Prop on end and allow to air dry	Clean dry surface

Procedure 13. How to Clean Pots, Pans, and Small Utensils (Manual)

When	How	Use
After each use	1. Place soiled pots and pans on left-hand drainboard (a) Place small utensils in one pot	
	2. Fill soak sink ¾ full	Hot water, approximately 120 F (49 C)
	3. Scrape food particles from pots and pans	Heavy spatula, garbage disposal
	4. Place pans into soak sink (a) Place small utensils in one pot; avoid bending or crushing beaters and wire whips	
	Caution: DO NOT PLACE KNIVES OR SHARP UTENSILS IN SINK; WASH AND RINSE IMMEDIATELY AND ALLOW TO AIR DRY ON CLEAN DRAINBOARD	
	(b) Allow pans to soak while preparing other 2 sinks	
	5. Measure detergent into wash sink; fill ¾ full	Detergent Hot water, approximately 120 F (49 C)

Procedure 13. (Continued)

When	How	Use
	6. Fill rinse sink ¾ full	Hot running water, 170 F (76.6 C) or above [a]
	7. Scrub pots and pans in soak sink; transfer to wash sink	Metal sponge, gong brush
	8. Scrub pots and pans	Hot detergent solution
	(a) Transfer clean pots and pans to rinse sink, open side up	
	(b) Do not allow pots and pans to nest	
	(c) Place small utensils in one pot to avoid damage and make removal easier	
	9. Allow pans to remain immersed for one minute	Hot running water, 170 F (76.6 C) or above
	10. Remove pots and pans, place open side down, and allow to air dry	Double S Hook / Clean drainboard
	(a) Do not allow pans to nest	
	11. Stack dry pans in appropriate place on clean pot and pan rack open side down	
	12. Return small utensils to appropriate drawer or hang on utensil rack	
	13. Change soak and wash water as often as necessary	Follow steps 2 and 5

[a] *Hot water booster tank or steam line in rinse sink may be required to maintain 170 F (76.6 C). If this temperature cannot be maintained, pots and pans will need to be sanitized with chemical sanitizer.*

Procedure 14. How to Clean Pots, Pans, and Small Utensils (Machine)

When	How	Use
After each use	1. Follow steps 1 through 4 on manual procedure	
	2. Fill pot and pan machine	Hot water
	(a) Open steam valve or turn on gas burner	
	3. Fill detergent dispenser	Detergent

(continued)

Procedure 14. (Continued)

When	How	Use
	4. Rack pots and pans (a) Put like pans together (b) Do not overload or allow pans to nest	Appropriate rack
	5. Check temperature on wash and rinse tank	
	6. Push start button	Wash 140 F (60 C) Final Rinse 180 F (82 C)
	7. Push rack into machine	Automatic wash-and-rinse cycle
	8. Remove rack from machine	
	9. Allow pans to air dry (a) Do not allow pans to nest	Clean drainboard
	10. Stack dry pans on clean pot and pan rack, open side down	
	11. Return small utensils to appropriate drawer or utensil rack	
	12. Change water in soak sink as often as necessary	
	13. Clean pot and pan machine as often as necessary	

Procedure 15. How to Wash Dishes, Glassware and Flatware (Flite-Type Machine) [a]

When	How	Use
Soiled Dish Station		
After each use	1. Fill dish machine	Hot water
	2. Turn on steam	
	3. Fill detergent dispenser and rinse dry dispenser	
	4. Scrape, prerinse, and stack dishes according to size and type	Prerinse unit, garbage disposal
	5. Rack cups and glasses open side down	Cup and glass racks
	6. Separate flatware according to type	
	7. Place flatware into perforated baskets with business side up; soak	Perforated flatware baskets Flatware soaking sink

Procedure 15. (Continued)

When	How	Use
	8. Check temperature on wash- and-rinse tank	Prerinse 120 F (49 C) Wash 150 F (65.6 C) Pumped rinse 160 F (71 C) Final rinse 180 F (82 C)
	9. Push start button on machine	Automatic wash and rinse
	10. Place glass racks on conveyor belt	
	11. Place cup racks on conveyor belt	
	12. Rack dishes on belt according to size and type	
	13. Place filled flatware baskets in cup rack	
	(a) Place rank on conveyor belt	
Clean dish Station		
	14. Inspect glasses for cleanliness (set aside soiled glassware and return to soiled-dish table)	
	15. Remove glass racks from belt	To a clean dolly
	16. Inspect cups for cleanliness; remove any soiled cups.	
	17. Remove cup rack from belt	To a clean dolly
	18. Remove dishes from belt, inspect for cleanliness, and stack according to size and type	Clean dish table or appropriate dish cart
	19. Remove flatware	
	(a) Inspect for cleanliness	
	20. Transfer clean flatware to clean perforated basket, business side down	Perforated flatware basket
	21. Send flatware through dish machine a second time	
	22. Remove flatware basket from belt	Clean dish table or appropriate flatware cart
	23. Transport dishes, glasses, and flatware to point of use	Cart

[a] *Dish washer could check water temperature, detergent, and rinse-dry dispensers frequently to be sure machine is being operated properly.*

Procedure 16. How to Clean Rotary Toaster

When	How	Use
After each use	1. Brush crumbs from front surface of toaster behind bread racks	Small stiff-bristled brush
	2. Remove slanted toast guide and tray below toaster	
	3. Brush dry crumbs from guide, tray, under the toaster, and surrounding table top	Clean dry cloth
	4. Wash toaster frame and area under toaster	Clean damp cloth dipped into hot machine detergent solution
	5. Wash guide, tray, and surrounding table top	Same as above
	6. Rinse	Clean hot water, chemical sanitizer, clean cloth
	7. Wipe dry	Clean dry cloth
	8. Replace toast guide and tray	

Procedure 17. How to Clean Automatic Toaster

When	How	Use
After each use	1. Remove electric plug from socket	
	2. Brush interior of each slot free of crumbs	Small stiff-bristled brush
	Caution: AVOID INJURY TO WIRES	
	3. Pull out crumb tray beneath toaster, brush free of crumbs	Clean dry cloth or soft brush
	4. Move toaster, brush crumbs from table top	Same as above
	5. Wash table top, toaster, and crumb tray	Clean damp cloth dipped in hot machine detergent solution
	6. Rinse	Clean hot water, chemical sanitizer, clean cloth
	7. Wipe dry	Clean dry cloth
	8. Replace toaster and crumb tray	
	9. Connect electric plug to socket	

Procedure 18. How to Clean Coffee Urn

When	How	Use
After each use	1. Lift cover 2. Remove bag or filter [a] (a) Empty grounds (b) Wash urn bag or filter (c) Soak bag when not in use 3. Drain urn 4. Close drain 5. Rinse urn and drain 6. Close drain; remake coffee as needed	 Hot running water Cold water Clean hot water, long-handled gong brush
Daily	1. Follow steps 1 through 5 above 2. Close drain and partially fill 3. Scrub urn and inside of cover 4. Drain urn 5. Flush urn with faucet open 6. Clean glass gauge (a) Remove nut at top of tube (b) Insert brush and plunge up and down (c) rinse gauge glass 7. Clean faucet and shank (a) Remove faucet from shank (b) Scrub inside of shank into urn (c) Scrub inside of faucet (d) Rinse faucet (e) Assemble faucet to shank (f) Rinse shank; partially fill urn, leave faucet opens 8. Close faucet 9. Partially fill urn with water, leave until the next service 10. Wipe outside of urn and cover Scrub drain and urn stand 11. Wipe dry 12. Leave urn open	 Hot water Long-handled gong brush Hot water Hot water, urn cleaner gauge brush Clean hot water, measuring cup with pouring lip Hot water, urn cleaner Gauge brush Clean hot water Clean hot water 2 gallons of water Clean damp cloth Hot machine detergent Clean dry cloth

(continued)

Procedure 18. (Continued)

When	How	Use
Two to three times a week	1. Follow steps 1 to 5 in "after each use" schedule 2. Close drain 3. Partially fill urn with water and urn cleaner and mix well 4. Allow solution to stand 30 to 45 minutes 5. Follow steps 3 to 12 in the "Daily" Schedule	Hot water, one package of urn cleaner, long-handled gong brush

[a] *Coffee grounds should be removed as soon as coffee is finished.*

Procedure 19. How to Clean Potato Peeler

When	How	Use
After each use	1. Flush out machine while machine is running [a] 2. Turn off machine, allow disc to stop	Clean cold water
Daily	3. Remove lid 4. Lift abrasive disc by sunken handles 5. Flush machine again if any peelings remain 6. Scrub inside and outside of machine, give special attention to crevices around door 7. Disassemble peel trap (a) Remove lid (b) Remove strainer (c) Remove stopper 8. Flush peel trap Scrub 9. Take parts to sink and scrub (a) Lid (b) Disc (c) Strainer (d) Stopper	Clean cold water Hot water, gong brush Cold water Hot water, gong brush Hot water, gong brush [b]

Procedure 19. (Continued)

When	How	Use
	10. Rinse parts and allow to air dry	Clean hot water
	11. Wipe exterior	Clean damp cloth
	12. Reassemble:	
	(a) Replace disc	
	(b) Replace stopper and strainer in peel trap	
	(c) Leave lid off peeler and peel trap to air	
	(d) Leave door open to air	

[a] *Always flush machine with cold water prior to using hot water. Hot water will cause peelings to cook onto metal and make cleaning more difficult.*
[b] *Do not use detergent on the abrasive disc; it is difficult to rinse off all the detergent.*

Procedure 20. How to Clean Steam-Jacketed Kettle

When	How	Use
After each use	1. Turn off steam, allow kettle to cool	
	2. Open drain	
	3. Flush kettle	Lukewarm water
	4. Close drain, remove drain screen [a]	Long-handled gong brush
	5. Fill kettle to cover highest food line, add detergent, mix well	Hot water, detergent / Long-handled gong brush
	6. Withdraw approximately 2 quarts of water to charge outlet pipe	Gallon measure
	7. For cooked-on soil, allow solution to stand for 30 to 60 minutes	
	8. Scrub kettle, inside and out, scrub cover inside and out, scrub frame and drain cover	Hot machine detergent solution, long-handled gong brush
	9. Open drain, remove valve core shell and outlet port	
	10. Rinse kettle and all parts	Clean hot water
	11. Reassemble outlet port	

(continued)

Procedure 20. (Continued)

When	How	Use
	12. Close drain	
	13. Sanitize outlet port, drain	Chemical sanitizer in proper solution
	14. Close drain, replace drain screen	
	15. Allow inside of kettle to air dry	
	16. Wipe outside of kettle, cover, and frame	Clean dry cloth
	17. Close cover	

a Newer models may not have drain covers.

Procedure 21. How to Clean Stack Steamer

When	How	Use
Daily	1. Turn off steam, allow to cool	
	2. Start cleaning with top deck	
	3. Turn center wheel counterclockwise to open	
	4. Remove shelves a	
	(a) Release safety catch under shelf	
	(b) Lift shelf up and out	
	5. Take shelves to pot-and-pan sink	
	6. Scrub top and bottom of shelves	Hot machine detergent solution, gong brush
	7. Rinse	Clean hot water, 170 F (76.6 C) for 1 minute
	8. Allow shelves to air dry on clean surface	
	9. Scrub chambers, check to see that all drains are open, scrub outside, sides, and top of steamer	Hot machine detergent solution, gong brush, gauge brush
	10. Rinse chambers, outside and top	Clean hot water, clean cloth
	11. Close steamer, turn on for 1 minute	

Procedure 21. (Continued)

When	How	Use
	12. Turn off steamer, open, allow to air dry	
	13. Replace shelves, fasten safety lock, test lock	
	14. Wipe off outside and top of steamer	Clean dry cloth
	15. Leave doors ajar	

a Newer models may have pan glide instead of shelves.

Procedure 22. How to Clean Hi-Pressure Steamer (Figure 19)

When	How	Use
Daily	1. Turn off steam	
	2. Allow to cool	
	3. Turn center wheel counterclockwise as far as it will go; push to right to open	
	4. Scrub inside and outside of chamber, give special attention to pan glides, clean drain	Hot machine detergent solution, gong brush, gauge brush
	5. Rinse chamber, inside and out	Clean hot water Clean cloth
	6. Close door, turn on steam for 1 minute	
	7. Open door, allow to air dry	
	8. Wipe outside and top	Clean dry cloth

Procedure 23. How to Clean Electric Overhead Broiler (Salamander)

When	How	Use
After each use	1. Turn off broiler, allow to cool	
	2. Remove grid:	
	(a) Pull lift lever to left	
	(b) Move grid to highest position	
	(c) Lift grid up and out	

(continued)

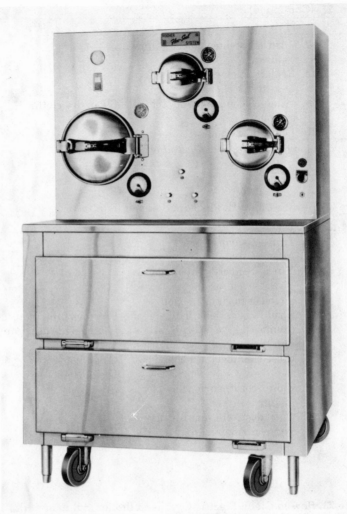

Figure 19. Hi-pressure steamer with freezer base. (Courtesy of Vischer Products Co.)

Procedure 23. (Continued)

When	How	Use
	3. Take grid to pot-and-pan sink, soak	Hot machine detergent solution
	4. Scrape any burnt particles from sides of broiler	Long-handled scraper
	5. Remove grease tray, empty grease	Waste-fat container
	6. Take grease tray to pot-and-pan sink	
	7. Scrub broiler grid and grease tray	Hot machine detergent solution, gong brush
	8. Rinse grid and grease tray, allow to air dry on clean surface	Clean hot water, 170 F (76.6 C) for 1 minute
	9. Wipe inside of broiler	Clean damp cloth wrung out in hot machine detergent solution
	Caution: DO NOT USE TOO MUCH WATER; DO NOT WASH CALROD UNITS—THEY ARE SELF-CLEANING	
	10. Wipe outside and top of broiler	Same as above
	11. Rinse inside, outside, and top of broiler, wipe dry	Clean damp cloth wrung out in clean hot water, clean dry cloth
	12. Replace grease tray	
	13. Replace grid	
	(a) Lock into position	
	(b) Test grid to see if it will roll out and lock	

Procedure 24. How to Clean Gas Frialator

When	How	Use
After each use	1. Allow fat to cool but not solidify	
	2. Drain and strain fat [a]	Cheese cloth, 2 or 3 thicknesses, clean pot
	(a) Secure cheesecloth to pot to form strainer	
	(b) Place pot below drain	
	(c) Open drain	
	(d) Close drain after all fat has been removed	
	(e) Move fat from frialator	

(continued)

Procedure 24. (Continued)

When	How	Use
	3. Fill frialator, scrub inside and outside of unit	Hot machine detergent solution, long-handled gong brush
	4. Place large pot below drain and open drain	Large pot
	5. Rinse frialator thoroughly	Clean hot water
	6. Wipe dry	Clean dry cloth
	7. Refill frialator [b]	Strained fat
	8. Cover when not in use	
Before changing to new fat	1. Allow fat to cool but not to solidify	
	2. Drain fat and dispose	Frozen fruit cans, 30 #
	3. Flush unit thoroughly to remove all sediment	Hot water, waste container
	4. Fill frialator	Warm water, detergent, or deep-fry cleaner
	5. Turn on frialator and boil for 30 minutes [c]	
	6. While boiling, scrub inside of the frialator	Long-handled gong brush, rubber gloves
	Caution: AVOID SPLASHING ANY SOLUTION ON SKIN OR IN EYES. IF ANY COMES IN CONTACT WITH SKIN, FLUSH IMMEDIATELY WITH LARGE QUANTITIES OF CLEAR WATER [d]	
	7. Drain frialator	Waste container
	8. Rinse unit thoroughly	Hot water
	9. Refill tank	Warm water, detergent
	10. Brush all inside surfaces, give special attention to burners and crevices under burners, clean outside	Long-handled brush
	11. Drain	Waste container
	12. Rinse	Hot water, ½ cup white vinegar
	13. Dry thoroughly	Clean dry cloth
	14. Fill with new fat	
	15. Cover when not in use	

[a] *Or use commercial fat-filter machine, and pump filtered shortening into clean pot.*
[b] *If frialator is not used daily, grease clean frialator lightly with clean shortening, cover, and store strained shortening in refrigerator.*
[c] *Put burner grids from range into this solution and boil—see procedure on page 131.*
[d] *Protective eye goggles may be required by law.*

Procedure 25. How to Clean Electric Frialator

When	How	Use
After each use	1. Allow fat to cool but not solidify	
	2. Lift calrod unit from fat, lock into position	
	3. Turn on frialator to high temperature (calrod units are self-cleaning) [a]	
	4. Take clean container from storage cabinet below unit; place on range or cart next to frialator	Clean frialator container
	5. Pump fat from frialator through filter into clean container	Commercial fat filter machine
	6. Remove soiled container from frialator, take to pot-and-pan sink for cleaning	Hot machine detergent solution, gong brush
	7. Rinse and allow to air dry	Clean hot water, 170 F (76.6 C) for 1 minute
	8. Turn off frialator, allow unit to cool	
	9. Scrub shell, surrounding frame and outside	Hot machine detergent solution, clean damp cloth
	10. Rinse	Clean hot water, chemical sanitizer, clean cloth
	11. Wipe dry	Clean dry cloth
	12. Lift clean container with strained fat into frialator [b]	
	13. Remove safety lock, lower calrod unit into fat	
	14. Cover when not in use	

[a] *If time permits, clean calrod units first, then proceed with remainder of schedule.*
[b] *If frialator is not used daily, cover and store strained fat in refrigerator.*

Procedure 26. How to Clean Electric Fry Pan (Figure 20)

When	How	Use
After each use	1. Turn off switch	
	2. Raise cover	
	3. Flush pan to remove all loose food particles	Lukewarm water, gong brush

(continued)

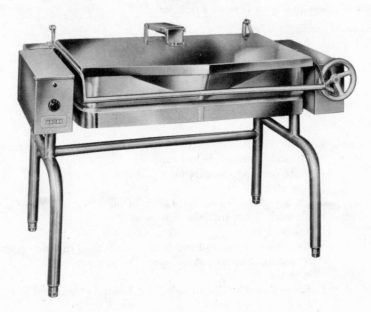

Figure 20. Electric fry pan. (Courtesy of Groen Mfg. Co.)

Procedure 26. (Continued)

When	How	Use
	4. Tilt pan to remove water (a) Turn wheel at right side of pan counterclockwise to pouring position [a]	Large pail or floor drain
	5. Return pan to upright position (a) Turn wheel clockwise	
	6. Fill fry pan to highest food line	Hot water, detergent
	7. Scrub pan, give special attention to cover and pouring lip; scrub outside of pan, cover, and frame	Gong brush
	8. Tilt pan to remove water (a) Turn wheel counterclockwise	Large pail or floor drain
	9. Rinse pan inside and out	Hose, hot water

Procedure 26. (Continued)

When	How	Use
	10. Return pan to upright position (a) Turn wheel clockwise	
	11. Allow inside to air dry; wipe outside of pan, cover, and frame dry	Clean dry cloth
	12. Close cover	

[a] *Place bun pan in front of pouring lip to guide water to floor drain and to prevent splashing.*

Procedure 27. How to Clean Ranges

When	How	Use
Grill Plates		
After each use	1. Allow plate to cool	
	2. Scrape grill to loosen burned-on particles	Heavy turner or putty knife
	3. Remove loose particles	Cloth, dustpan
	4. Scrub grill plate	Hot machine detergent solution, gong brush
	5. Scrub grease trough	Same as above
	6. Scrub back apron	Same as above
	7. Remove, dump, and wash grease trap	Same as above
	8. Rinse and wipe dry	Clean hot water Clean dry cloth
Open Top		
Daily	1. Allow range to cool	
	2. Lift out burner grids, take to sink and scrub	Hot machine detergent solution, gong brush Wire brush
	3. Rinse, wipe dry	Clean hot water, clean dry cloth
	4. Brush burners, light burners to check for clogged burners	Wire brush, ice pick
	5. Clean back apron and warming oven	Hot machine detergent, clean damp cloth
	6. Remove grease and/or warming trays below burners; take to sink and scrub	Same as above
	7. Rinse and wipe dry.	Clean hot water, clean dry cloth
	8. Replace clean grids and trays	

(continued)

Procedure 27. (Continued)

When	How	Use
Solid Top		
Daily	1. Allow range to cool	
	2. Loosen all burned-on particles	Putty knife, dustpan
	3. Scrub top of range (keep water to minimum)	Hot machine detergent solution, gong brush, clean damp cloth
	4. Clean back apron and warming oven	Same as above
	5. Rinse and wipe dry	Clean hot water, clean dry cloth
	6. Apply rust preventative to solid tops	Salad oil or clean shortening
Outside of All Ranges		
Daily	1. Scrub fronts, sides, and back	Hot machine detergent solution, clean damp cloth
	2. Rinse	Clean hot water
	3. Wipe dry	Clean dry cloth
	4. Polish all metal trim	Metal cleaner, clean soft cloth
Solid Top		
Weekly	1. Allow range to cool	
	2. Remove center lid	Ring hook
	3. Remove ring and side plates	Hand
	4. Spread paper on firebrick	Brown paper or newspaper
	5. Scrape off burned-on particles from lid, ring, and side plates	Paper, putty knife
	6. Scrape edges and top surface	Putty knife, wire brush
	7. Soak parts 20 to 30 minutes	Hot machine detergent solution
	8. Scrub parts, edges, and top surface	Hot machine detergent solution, gong brush
	9. Rinse, wipe dry, remove paper from firebrick	Clean hot water, clean dry cloth
	10. Reassemble all parts of the top	Hand and ring hook
	11. Apply rust preventative	Salad oil or clean shortening

Procedure 27. (Continued)

When	How	Use
Burner Grids		
Each time frialator shortening is changed	1. Remove burner grids 2. Boil burner grids while cleaning frialator 3. Take to pot-and-pan sink, scrub 4. Rinse and wipe dry 5. Replace burner grids	Frialator solution Hot machine detergent solution, gong brush, wire brush Clean hot water, clean dry cloth

Procedure 28. How to Clean Range Ovens

When	How	Use
Daily	1. Allow oven to cool 2. Remove shelf 3. Scrape burned particles from hearth 4. Brush out interior, shelf ledges, and door crevice 5. Replace shelf 6. Wash outside of door and frame 7. Rinse and wipe dry	 Long-handled metal scraper Long-handled brush Cloth wrung out in hot machine detergent solution Clean hot water, clean dry cloth
Weekly	1. Allow oven to cool 2. Remove shelf, take to pot-and pan-sink, and scrub 3. Rinse and wipe dry 4. Scrape burned-on particles from hearth 5. Brush interior, shelf ledges, and door crevices 6. Scrub interior, shelf ledges, inside and outside of door, and frame 7. Rinse inside and outside of oven 8. Replace clean shelf 9. Wipe outside dry and polish metal trim	 Hot machine detergent solution, gong brush, metal sponge Clean hot water Clean dry cloth Long-handled metal scraper Long-handled brush, dustpan Hot machine detergent solution, long-handled brush Clean hot water, clean damp cloth Clean dry cloth Metal polish

Procedure 29. How to Clean Electric Stack Ovens

When	How	Use
Daily	1. Allow oven to cool	
	2. Scrape each deck to loosen all burned-on particles	Long-handled metal scraper
	3. Brush burned particles from each deck, starting with top deck. Give special attention to door crevice. Check to see if oven door closes properly	Long handled brush and dustpan
	4. Scrub inside and outside of each oven door	Metal sponge, gong brush, hot machine detergent solution
	5. Wash outside of oven, start with top, do sides, back, and front	Same as above
	6. Rinse and wipe dry	Clean hot water Clean dry cloth
	7. Polish any metal surfaces, handles or trim	Appropriate metal cleaner Clean soft cloth

Procedure 30. How to Clean Gas Stack Ovens

When	How	Use
Daily	1. Allow oven to cool	
	2. Scrape each oven deck to loosen burned-on particles	Long-handled metal scraper
	3. Brush burned particles from each deck, starting with top deck. Give special attention to door crevices. Check to see if oven doors close properly	Long-handled brush and dustpan
	4. Brush out burner compartments. Light burners, check for clogged burners, if so, remove and clean	Radiator type brush Ice pick
	5. Scrub inside and outside of each oven door	Metal sponge, gong brush, hot machine detergent solution
	6. Wash outside of oven, start with top, do sides, back and front	Same as above
	7. Rinse, wipe dry	Clean hot water Clean dry cloth
	8. Polish any metal surfaces, handles, or trim	Appropriate metal cleaner and soft cloth

Procedure 31. How to Clean Pot-and-Pan Machine

When	How	Use
After each use	1. Close steam valve or turn off gas burner	
	2. Lift overflow funnel to drain tank	
	3. Remove and empty filter screens	
	4. Scrub screens and overflow funnel in soak sink	Hot machine detergent solution, gong brush
	5. Flush interior of machine, scrub inside while flushing machine	Hose, hot water, long-handled gong brush
	6. Hose down inside of machine (this cools metal and makes it easier to remove stains)	Cold water, hose
	7. Replace overflow funnel and screens	
	8. Wipe off outside and top of machine	Hot machine detergent solution, clean cloth
	9. Rinse, wipe dry	Clean water, clean dry cloth
	10. Scrub drainboards	Hot machine detergent solution, gong brush
	11. Rinse, wipe dry	Clean water, chemical sanitizer, clean dry cloth
	12. Leave doors open when not in use	
Weekly	1. Follows steps 1 to 7 in above schedule	
	2. Fill machine	Hot water, detergent
	3. Open steam valve or turn on gas burner	
	4. Start machine, allow machine to run for 30 minutes	
	5. Turn off steam valve or gas burner	
	6. Remove filter screens	
	7. Remove overflow funnel to drain machine	
	8. Scrub inside of machine while draining	Long-handled gong brush
	9. Flush machine	Cold water
	10. Replace overflow funnel and filter screens	

(continued)

Procedure 31. (Continued)

When	How	Use
	11. Wipe down sides and top of machine, scrub drainboards	Hot machine detergent solution, gong brush, clean damp cloth
	12. Rinse, wipe dry	Clean water, chemical sanitizer, clean dry cloth
	13. Leave doors open when not in use	

Procedure 32. How to Clean Flite-Type Dishmachine

When	How	Use
After each meal	1. Pull scrap trays and curtains	
	2. Clean trays over garbage disposal	Hot water, hose, gong brush
	3. Run trays and curtains through dish machine	
	4. Turn off dishmachine	
	5. Scrub down dish tables	Hose, hot water, long-handled gong brush
	6. Rinse and squeegee dish tables	Squeegee
	7. Drain machine; scrub tanks with brush while draining	Long-handled gong brush
	8. Hose down inside of machine (this cools metal and makes it easier to remove stains)	Hose, cold water
	9. Wipe soap dispenser electrodes, exterior of dispenser and machine	Clean damp cloth
	10. Wipe dry	Clean dry cloth
	11. Fill tanks and allow water to overflow for 2 to 3 minutes	
	12. Replace scrap trays, dispenser pan, and curtains	
After evening or last meal	1. Do not fill tanks	
	2. Do not replace curtains until next use	
Weekly	1. Clean machine the same as after each meal	
	2. Remove wash arms and clean free of foreign material	Long-handled bottle brush, ice pick

Procedure 32. (Continued)

When	How	Use
	3. Replace wash arms	
	4. Remove rinse jets	
	(a) Remove lime or iron deposits	
	(b) Tie rinse jets in clean cloth and place in machine	
	5. Take stainless steel equipment such as water fountain, splash troughs and covers, and place in machine	
	6. Turn off dispenser and final rinse	
	7. Fill machine and add detergent	Appropriate amount of detergent for each tank
	8. Run machine for 45 minutes	
	9. Clean dispenser electrodes	Damp cloth
	10. Drain machine, scrub tank while draining	Long-handled gong brush
	11. Hose down machine thoroughly	Hose, cold water
	12. Replace rinse jets	
	13. Wipe other stainless steel equipment dry and replace	Clean dry cloth
	14. Clean exterior of machine and dispenser	Damp cloth, stainless cleaner
	15. Rinse, wipe dry	Clean water, clean dry cloth
	16. Turn on dispenser and final rinse	

Procedure 33. How to Clean Ice-Cream Cabinet

When	How	Use
After each meal	1. Cover ice cream	Carton lid or clean wax paper or film
	2. Wipe inside and outside of cabinet lid and rim of cabinet openings	Hot machine detergent solution, clean cloth
	3. Wipe front, sides, and top of cabinet	Same as above

(continued)

Procedure 33. (Continued)

When	How	Use
	4. Rinse and wipe dry	Clean hot water, chemical sanitizer, damp cloth, and clean dry cloth
	5. Clean dipper well	
	(a) Remove ice cream dippers and send to dishwasher	
	(b) Drain well, turn off running water	
	(c) Wash well	Hot machine detergent solution, damp cloth
	(d) Turn on water, rinse well	
	(e) Allow water to run slowly	
	(f) Replace clean dippers	
Weekly	1. Cover ice cream	Carton lid or clean wax paper or film
	2. Remove ice cream to freezer	
	3. Disconnect plug from socket	
	4. Defrost freezer [a]	
	(a) Remove lids	
	(b) Place pan of hot water in each compartment	Very hot water, pans
	(c) Replace lids	
	(d) Steam for 10 to 15 minutes	
	(e) Replace water with more hot water, if necessary	
	5. Remove loosened sheets of ice or frost	Heavy spatula, pan
	6. Scrub interior, lids, rim openings, and outside of freezer	Hot machine detergent solution, gong brush, cloth
	7. Wipe up excess water	Clean dry cloth or sponge
	8. Rinse and repeat step 7	Clean hot water, chemical sanitizer
	9. Wipe lids, rim openings, and outside of freezer	Clean dry cloth
	10. Connect plug to power source	
	11. Replace ice cream when temperature reaches 10 F	
	12. Replace lids	

[a] *Steam hose may be used in place of hot water for defrosting.*

Procedure 34. How to Clean Reach-in Refrigerator

When	How	Use
Daily	1. Check contents for any item that needs to be covered or discarded	
	2. Wipe up spills on shelves, sides, and floor	Hot machine detergent solution, clean cloth
	3. Wash doors inside and out, door frames, and front	Same as above
	4. Rinse	Clean hot water, chemical sanitizer, clean cloth
	5. Wipe outside dry	Clean dry cloth
	6. Polish outside (if stainless steel)	Stainless steel polish Soft dry cloth
Weekly	1. Transfer all items to walk-in refrigerator	Cart
	2. Remove shelves, take to pot-and-pan sink [a]	Cart
	3. Scrub shelves top and bottom	Hot machine detergent solution, gong brush
	4. Rinse	Clean hot water, 170 F (76.6 C) for 1 minute
	5. Air dry	Clean surface
	6. Scrub box inside and out give special attention to: (a) Shelf glides (b) Door frame and hinge area	Hot machine detergent solution, gong brush, clean cloth
	7. Rinse	Clean hot water, clean cloth
	8. Sanitize, allow to air dry	Cloth saturated with sanitizing solution
	9. Replace shelves	
	10. Wipe outside of box dry	Clean dry cloth
	11. Polish (if stainless steel)	Stainless steel polish, soft dry cloth
	12. Replace food items as soon as box temperature reaches 40 F	

[a] *Shelves may be cleaned in dish machine or pot-and-pan machine.*

Procedure 35. How to Clean Steam Table

When	How	Use
After each use	1. Turn off steam 2. Drain 3. Wash steam-table adapters [a] surrounding frame, and overhead shelf	Hot machine detergent solution, clean cloth
	4. Rinse	Clean hot water, chemical sanitizer, clean cloth
	5. Wipe dry and replace adaptors	Clean dry cloth
	6. Wash front and back of steam table, wipe dry	Hot machine detergent solution, clean dry cloth
	7. Polish stainless steel as needed	Stainless steel polish, soft dry cloth
Daily after evening meal	1. Turn off steam 2. Add detergent to water in steam table	Machine detergent
	3. Wash steam table adaptors, surrounding frame, and overhead shelf	Hot machine detergent solution, clean cloth
	4. Rinse adaptors and overhead shelf, wipe dry; lay adaptors aside	Clean hot water, chemical sanitizer, clean cloth Clean dry cloth
	5. Scrub inside of steam table, give special attention to steam pipes and crevice below pipes	Long-handled gong brush
	6. Drain	
	7. Rinse	Clean water
	8. Wipe surrounding frame dry	Clean dry cloth
	9. Replace adaptors and covers	
	10. Wash front and back of steam table	Hot machine detergent solution, clean cloth
	11. Rinse	Clean hot water, chemical sanitizer, clean cloth
	12. Wipe dry	Clean dry cloth
	13. Polish stainless steel as needed	Stainless steel polish, soft dry cloth

[a] Adaptors may also be washed in dish machine.

Procedure 36. How to Clean Dish Trucks

When	How	Use
After each use	1. Wipe up any spilled material	Cloth saturated with hot machine detergent solution
	2. Rinse	Clean warm water
	3. Wipe dry	Clean dry cloth
Daily	1. Scrub top, shelves, frame	Hot machine detergent solution, gong brush
	2. Rinse	Clean warm water
	3. Sanitize top and each shelf	Cloth saturated with chemical sanitizer
	4. Air dry	
	5. Polish frames and handle	Clean dry cloth

Procedure 37. How to Clean Baker's Table

When	How	Use
Top and Sides		
As needed throughout the day and at end of day	1. Scrape table top	Bench knife
	2. Remove all scrapings and dry flour	Dry cloth, pan
	3. Scrub top, sides and spice shelf	Hot machine detergent solution, gong brush
	4. Rinse	Clean hot water, clean cloth
	5. Sanitize top	Clean cloth saturated with chemical sanitizer solution
	6. Air dry top, wipe sides and shelf dry	Clean dry cloth
Drawers		
Weekly	1. Remove all utensils	Clean tray
	2. Remove drawers from table (a) Press in on safety clamp	
	3. Take drawers to pot-and-pan sink	Cart
	4. Scrub drawers, inside and outside	Hot machine detergent solution, gong brush
	5. Rinse	Clean hot water, 170 F (76.6 C) for 1 minute
	6. Dry	Clean dry cloth
	7. Wipe drawer frames	Clean damp cloth dipped into hot machine detergent solution

(continued)

Procedure 37. (Continued)

When	How	Use
	8. Rinse and wipe dry	Clean hot water, chemical sanitizer, clean dry cloth
	9. Return drawers to table	Cart
	(a) Press in safety clamp to insert	
	(b) Release safety clamp to lock	
	(c) Test lock	
	10. Put clean utensils back into drawers	
Table Bins		
Before refilling	1. Remove bins from table	
	(a) Press in on safety clamp	
	2. Empty any remaining material into clean container	Clean container
	3. Vacuum each bin cavity and inside frame to remove all dry flour or sugar	Vacuum cleaner
	4. Take bins to pot-and-pan sink	Pallet truck
	5. Scrub bins, inside and outside	Hot machine detergent solution, gong brush
	6. Rinse	Clean hot water, 170 F (76.6 C) for 1 minute
	7. Wipe dry	Clean dry cloth
	8. Return to baker's table	Pallet truck
	9. Wash inside frame of table	Damp clean cloth dipped into hot machine detergent solution
	10. Rinse	Clean hot water, chemical sanitizer, clean cloth
	11. Wipe dry	Clean dry cloth
	12. Replace bins	
	(a) Press in on safety clamp to insert	
	(b) Release safety clamp to lock	
	(c) Test lock	
	13. Refill, put new material in first	

Procedure 38. How to Clean Kitchen Tables

When	How	Use
Table Tops		
As needed throughout the day and at the close of each shift	1. Remove any loose food particles	Clean damp cloth
	2. Scrub	Hot machine detergent solution, gong brush
	3. Rinse	Clean hot water, clean cloth
	4. Sanitize	Clean cloth saturated with chemical sanitizer solution
	5. Allow to air dry	
Drawers		
Weekly	1. Remove all utensils	Clean container
	2. Remove drawers from table	
	(a) Press in on safety clamp	
	3. Take drawers to pot-and-pan sink	Cart
	4. Scrub drawers inside and out	Hot machine detergent solution, gong brush
	5. Rinse	Clean hot water, 170 F (76.6 C) for 1 minute
	6. Wipe dry	Clean dry cloth
	7. Wipe drawer frames	Clean damp cloth dipped into hot machine detergent solution
	8. Rinse and wipe dry	Clean hot water, chemical sanitizer, clean dry cloth
	9. Return drawers to table	Cart
	(a) Press in safety clamp to insert	
	(b) Release safety clamp to lock	
	(c) Test lock	
	10. Put clean utensils back into drawers	
Shelf under Table		
Weekly	1. Remove everything from shelf	
	2. Scrub shelf	Hot machine detergent solution, gong brush
	3. Rinse	Clean hot water, chemical sanitizer, clean cloth
	4. Wipe dry	Clean dry cloth
	5. Replace matrials on shelf	

Procedure 39. How to Clean Metal Shelving (Pot-and-Pan Storage) [a]

When	How	Use
Weekly	1. Remove pots and pans to clean surface, place upright	Sink, drainboard, or cart
	2. Scrub top and bottom of metal shelving	Long-handled gong brush Hot machine detergent solution
	3. Rinse	Clean warm water
	4. Sanitize	Clean cloth saturated with chemical sanitizer solution
	5. Allow to air dry	
	6. Replace pots and pans, top down	

[a] *Use same procedure for cleaning all metal shelving.*

Procedure 40. How to Clean Microwave Oven

When	How	Use
After each use	1. Wipe up spills	Clean damp cloth dipped into mild detergent solution
	2. Rinse and wipe dry	Clean damp cloth dipped into clean warm water, chemical sanitizer, clean dry soft cloth
Interior		
Daily	1. Remove glass cooking tray and temperature probe [a]	
	2. Wipe out cavity and door *Caution:* BE SURE OVEN FACE AND DOOR ARE KEPT CLEAN AT ALL TIMES TO INSURE POSITIVE DOOR SEAL	Clean damp cloth dipped into warm, mild detergent solution
	3. Rinse and wipe dry	Clean damp cloth dipped into warm water, chemical sanitizer, clean dry cloth
	4. Wash glass cooking tray and temperature probe [b]	Clean damp cloth dipped into mild detergent solution
	5. Rinse glass cooking tray and temperature probe	Clean damp cloth dipped into warm water, chemical sanitizer
	6. Dry glass cooking tray and temperature probe and replace into oven	Clean soft cloth

Procedure 40. (Continued)

When	How	Use
Exterior		
Daily	1. Wipe exterior of oven and door *Caution:* BE SURE OVEN FACE AND DOOR ARE KEPT CLEAN AT ALL TIMES TO INSURE POSITIVE DOOR SEAL	Clean damp cloth dipped into warm mild detergent solution
	2. Rinse	Clean damp cloth dipped into clean warm water, chemical sanitizer
	3. Wipe dry	Clean soft cloth

[a] *Some models do not have glass cooking tray and temperature probe.*
[b] *Some models do not have glass cooking tray and temperature probe, therefore omit steps 4, 5, 6.*

Cleaning cloths or sponges are necessary not only for cleaning equipment but also to wipe up food spills during the hours of operation. If not handled correctly, these materials are a potential source of contamination. The foodservice worker should be trained in the proper maintenance of cleaning/wiping cloths or sponges so they will be readily available when needed and yet safe to use. Recommendations for handling these materials as given in section 5-102 of the *Food Service Sanitation Manual 1976* are as follows:

- Cloths used for wiping food spills on tableware, such as plate, bowls, or cups being used to serve food to the consumer, should be clean, dry, and used for no other purpose.
- Moist cloths or sponges used for wiping food spills on kitchenware and food-contact surfaces of equipment should be clean and rinsed frequently in one of the sanitizing solutions permitted in section 5-103 of the *Food Service Sanitation Manual 1976* and used for no other purpose. These cloths or sponges should be stored in the sanitizing solution betwen uses.
- Moist cloths or sponges used for cleaning nonfood contact surfaces of equipment such as counters, dining tabletops, and shelves should be cleaned and rinsed as specified above and used for no other purpose. These cloths or sponges should be stored in sanitizing solution between use.

SECTION C

Safety and Safety Regulations

Safety is an integral part of sanitation. The truth of that statement can best be documented by observing the "CAUTION" statements that appear in many of the cleaning procedures in Section B. According to the National Safety Council, the frequency rates of accidents in the foodservice industry are nearly twice as high as the average for all industries reporting. However, the severity rates for the industry fall about midway in all industry classification. "Accidents don't happen; they are caused," therefore they can be prevented—but only through the concerted effort of management and the workers.

Safe and healthful working conditions are mandated by the Williams-Steiger Occupational Safety and Health Act of 1970 which became effective April, 1971. This Act, better known as OSHA, provides safety and health protection for workers. The U.S. Department of Labor has primary responsibility for administering the Act. The department issues job safety and health standards, and employers and employees are required to comply with these standards.

BY LAW: SAFETY ON THE JOB IS EVERYBODY'S RESPONSIBILITY!

OSHA requires that each employer furnish employees a place of employment free from recognized hazards that might cause serious injury or death, and comply with the specific safety and health standards issued by the Department of Labor. OSHA also requires that each employee comply with safety and health standards, rules regulations, and orders issued under the Act and applicable to his or her conduct.

OSHA allows a compliance officer to enter the place of business to check for adherence to standards and to be sure the work place is free of recognized hazards. The compliance officer may enter on a general inspection or on a specific complaint. The officer will come unannounced. The place of business will be inspected under any of the following circumstances:

1. If a catastrophe or fatal accident has occurred in the foodservice operation.
2. When OSHA has received a valid employee complaint.
3. If there is an apparent imminent danger. (This is usually not applicable to the foodservice industry).
4. When the operation has been randomly selected for monitoring or statistical purposes by either federal or state officials.

There are four types of violations normally considered on the first inspection.

1. *De minimus:* A condition that has not direct or immediate relationship to job safety and health, for example, lack of toilet partitions.
2. *Nonserious violations:* A violation that does have a direct relationship to job safety and health but probably would not cause death or serious physical harm, for example, tripping hazard. A proposed penalty of up to $1000 is optional.*
3. *Serious violation:* A violation where there is substantial probability that death or serious physical harm could result and that the employer knew, or should have known of the hazard, for example, absence of operation guards on food choppers, slicers, and/or grinders. A proposed penalty of up to $1000 is mandatory.
4. *Imminent danger:* A condition where there is reasonable certainty that a hazard exists that can be expected to cause death or serious physical harm immediately or before the hazard can be eliminated through regular procedures (no example for foodservice is given). If the employer fails to abate such conditions immediately, the compliance officer, through his or her area director, can go directly to the nearest federal district court for legal action as necessary.

The inspection is conducted by an occupational safety and health compliance officer, who must display proper credentials upon entry. The officer should inform management of the reason for the visit and outline in general terms the scope of the inspection. Some of the items that may be checked during the inspection are:

1. The accessibility of fire extinguishers, and their readiness or use.
2. Guards on floor openings, balcony storage areas, and receiving docks.
3. Proper and adequate handrailings on all stairs.
4. Guards for machines and machine parts.
5. Electrical grounding of equipment and tools.

*Penalties may be adjusted downward depending on severity of the hazards, the employer's good faith, history of previous violations, and the size of the business.

6. General housekeeping with areas clean and in good repair.
7. Passageways lighted and clear of obstruction.
8. Compliance with OSHA posting requirements.
9. Compliance with OSHA record keeping.
10. Proper use of extension cords.
11. Exits marked and clear.

In addition, any recognized hazard of a serious nature not covered by a standard may be cited under the general duty clause.

After the inspection (walk-around), the compliance officer will hold a closing conference with management to review probable violations. Management will be asked to estimate the amount of time that will be needed to abate the hazard and the cost involved. The compliance officer may *not* personally impose or propose a penalty on the spot at an inspection, nor can he or she close down the place of business. The officer must return to his office, write a report and discuss it with the area director. The area director or the director's superiors determine what citations will be issued and what penalties will be proposed (if any). Citations are sent to management by certified mail, with a copy to the complaining employee if there is one. Citations may be contested but they are subject to final action by the Occupational Safety and Health Review Commission.

RECORD KEEPING AND POSTING

OSHA requires employers of eleven or more employees to keep certain records of job-related fatalities, injuries, and illnesses. In areas where state plans are in effect, the basic reporting is identical, but states may add additional items.

There are two forms required, neither of which must be submitted to OSHA. However, they must be made available when a compliance officer makes an inspection.

1. *OSHA No. 200–Log and Summary of Occupational Injuries and Illnesses:* Each employer who is subject to the recordkeeping requirements of OSHA must maintain for each establishment a log of all recordable occupational injuries and illnesses. A copy of the totals and information following the fold line of the last page for the year must be posted at each establishment, in the place where notices to employees are customarily posted, not later than February 1 and must remain in place until March 1 each year. Logs must be maintained and retained for five years following the end of the calendar year to which they relate. Logs must be available for inspection and copying by representatives of the Department of Labor.

2. *OSHA No. 101–the Supplementary Record:* For every recordable injury or illness, it is necessary to record additional information requested on OSHA No. 101 form. Worker's compensation, insurance, or other reports are acceptable supplementary records if they contain all items found on the OSHA No. 101 form.

All record forms must be maintained in each establishment for five years following the end of the year to which they relate. The law requires that employees be informed of the job safety and health protection provided under the Act.* This is attained by requiring management to post the "OSHA Poster" in a prominent place in the establishment to which the employees usually report to work.

Forms and posters may be obtained from the Office of the Assistant Secretary of Labor for Occupational Safety and Health, 200 Constitution Ave., N.W., Washington D.C. 20210 or from any of the regional offices of the Bureau of Labor Statistics or the Occupational Safety and Health Administration. For more detailed information about OSHA as it relates to food service operations, see *OSHA Reference for Food Administrators.* Copies may be secured from the American Hospital Association, 840 North Lake Shore Drive, Chicago, Ill. 60611.

*See Appendix.

SECTION D

A Sound Food Supply: Purchase and Sanitary Storage

In order for a sanitation program to be effective, it should start with a sound food supply. The person entrusted with purchasing is responsible for getting the best possible products for the money spent. To assure wholesomeness, it is of greatest importance to specify exactly what one wants. Certain government agencies concerned with the protection of our food supply are doing an excellent job for us in connection with safeguarding the wholesomeness of food. This aid is given on various levels of government—federal, state, and local. However, the final responsibility for serving wholesome menu items or meals to the public rests with the management of a foodservice establishment.

First, the food buyer should take no chances when selecting vendors. A buyer should only deal with reputable suppliers whose standards of sanitation are high and who use approved techniques in the handling of their products. High sanitary standards need to be followed regarding processing plants, warehouses, packaging, and vehicles used for transporting food items. Whoever is entrusted with purchasing food for an establishment should feel obligated to personally inspect the vendors' premises to be personally assured of the sanitary techniques applied here. Also, an inquiry should be made into whether microbiological criteria are set up for potentially hazardous products: what these are and if they are maintained at an appropriate level. For more information on microbiological criteria, consult Longrée, K., "Quantity Food Sanitation."*

The use of preprocessed (convenience) foods is, in some ways, making the problem of practicing good food sanitation increasingly more complex because the supplier must be relied upon to do his or her share in the production of high-quality food that is bacteriologically safe. Beware of vendors who refuse to admit a food buyer to their premises. Large foodservice companies (chains) even employ their own inspectors who check up on

*See the Readings for Part Three.

the maintenance of quality in the production of the items they are contracting for. Food-processing plants and other related activities such as warehouse operations and maintenance of appropriate temperatures in trucks used for transporting perishable food items are scrutinized. Keeping frozen and chilled foods at appropriate low temperatures is a prerequisite for keeping them microbiologically safe, besides ensuring the maintenance of other quality characteristics.

AGENCIES CONCERNED WITH FOOD PROTECTION

Food protection is the concern of certain governmental agencies on the federal, state, and local levels. The work of these agencies is of tremendous importance in regard to the microbiological quality of our food supply since they have developed inspection programs for potentially hazardous food items and regulations for their processing and shipment. The foods for which U.S. inspection programs for production and processing are available are dairy products, meat, poultry, eggs, fish, and shellfish. Federal inspection programs are also available for the processing of frozen and canned foods. Many state and local agencies have adopted U.S. regulations and quality standards "as is" or with some changes, making them stricter.

Two departments of the federal government, the Department of Health and Human Services (formerly DHEW) and the Department of Agriculture, are of invaluable aid in matters of food protection. A third agency, the Department of Commerce, is concerned with one commodity: fish.

The Department of Health and Human Services (DHHS)

This important agency works to protect and advance the health of the American people. The *Public Health Service* (PHS) of the DHHS is subdivided into several components, one of which is greatly concerned with food safety, others less so. Two components of interest in connection with food sanitation are the Food and Drug Administration and the Center for Disease Control.

The Food and Drug Administration (FDA).   The agency's responsibilities are many, and food is only one of them. The FDA helps to protect people from unsafe and unpure foods, unsafe drugs and cosmetics, and many other potential hazards. The FDA enforces the Food, Drug and Cosmetic Act and is responsible for the purity and safety of food shipped in interstate (from state to state) commerce. The FDA polices the purity, quality, and labeling of foods by periodically inspecting food processing and storage facilities, taking samples, and analyzing them for microorganisms, foreign bodies, additives, and pesticides. It may request recall of foods if suspected injurious to health. Besides processed foods, the FDA is also concerned with the safety of shellfish. The agency also has responsibility for the safety of food and drinking water aboard planes, trains, buses, and ships operating under the U.S. flag.

FDA's activities are directed from the Bureau of Food. The Bureau:

1. Conducts research and develops standards and policy on the composition, quality, safety, and nutrition of foods and of food additives, colors, and other potentially harmful products.
2. Conducts research for the improvement of methods to detect, prevent, and control food contaminants that may be harmful to humans.
3. Plans, coordinates, and evaluates FDA's surveillance and compliance programs concerning foods.
4. Reviews industry petitions, recommends the publication of regulations for food standards, permits the safe use of color additives and environmental factors affecting the total chemical hazard posed by food additives.
5. Maintains a nutritional data bank.

The FDA is also instrumental in giving guidance to state agencies in improving sanitation in foodservice operations. In fact, its *Food Service Sanitation Manual* (including a model foodservice sanitation ordinance) is indeed a model instrument for inspectors checking up on the sanitary condition in foodservice establishments all over the United States.

The Center for Disease Control.   This center, based in Atlanta, Georgia, has as its main responsibility to combat communicable diseases. However, it also makes an important contribution to food sanitation through its work with food-poisoning pathogens, such as the salmonellae. The Center also trains state and local health officials in epidemic control and operates local disease-prevention programs.

The Department of Agriculture (USDA)

Through one of its branches, the Food Safety and Quality Service, the USDA is responsible for the quality and safety of meat, poultry, and dairy products only; all other foods are under the control of the FDA. It enforces the Meat Inspection Act and the Poultry Products Inspection Act.

When meat and poultry are to be shipped across state lines, and also upon request, federal agents inspect the livestock and poultry before and after slaughter to ascertain that the meat products come from healthy animals and are wholesome. The inspectors also keep a watchful eye on sanitary procedures used in the packing plants.

On the state and local levels we have definite regulations that safeguard the wholesomeness of various hazardous and highly perishable food products. The food purchaser, whether the proprietor or manager, should be familiar with these regulations.

PURCHASING POTENTIALLY HAZARDOUS FOODS

Some foods must be considered potentially hazardous because of their composition and/or the handling they receive. We will now discuss the particular hazards one should be alerted to when purchasing these food groups; how and for how long one should store them; what the hazards of prolonged storage are; what the signs of quality loss are; and what one should do about suspect food. We will discuss milk and other dairy products, eggs, meat and poultry, seafood, and processed foods.

Milk and Other Dairy Products

The composition, sanitary production, and marketing of milk is regulated by strict federal, state, and local laws. If these are properly enforced, the public can be assured of receiving a high-quality product. Only a small percentage of all foodborne illnesses reported are caused by milk (see Figure 14). *Consult the state or local health departments for regulations that apply to the purchase and the service of milk in your foodservice operation.*

In properly pasteurized milk and cream, all pathogens and the majority of other microorganisms have been killed. Pasteurization involves heating the milk to specified temperatures, holding it there for specified times, and cooling it quickly. In the holding method, the milk is heated at a temperature of 145 F (63 C) for 30 minutes; with the high-temperature-short-time (HTST) method, the product is heated to at least 161 F (72 C) and held there for at least 15 seconds.

In ultrapasteurization (UHT) the milk is exposed to at least 280 F (137.8 C) for 2 seconds or more. This milk has a shelf life of several weeks when refrigerated. Unrefrigerated, its shelf life is shorter, but still much longer than that of milk pasteurized by the more common methods. The UHT method is, at present not popular in the U.S.A., but is used widely in some other countries. Whatever the heating method is, the milk must be cooled quickly to 45 F (7.2 C) or below.

Milk to be served as a beverage should for sanitary reasons be purchased in individual containers filled at the dairy; and served in individual, unopened containers, with the quantity not exceeding 1 pint (473 cubic cm), or drawn from a commercially filled container stored in a mechanically refrigerated bulk milk dispenser. In situations where a bulk dispenser, usually of 3 or 6 gallon (11.4 or 22.8 liter) capacity, is not available, and milk and milk products must be dispensed in quantities of less than ½ pint (236.5 cubic cm) for use with items such as cereal, mixed drinks, or desserts, milk and milk products may be poured from a commercially filled container whose capacity should not exceed ½ gallon (1.9 liters)*.

Whether or not milk may be dispensed for beverage purposes is a matter of local law. The manager should be informed about this matter. *If dispensing is permitted, detailed regulations must be followed regarding the care of the containers and contents, and serving procedures.* Refrigerated milk dispensers are recommended for dispensing bulk milk in kitchen and on serving line.

Milk drinks and frozen desserts like ice cream, sherbets, ice milks, and ices and mixes for frozen desserts should conform to sanitary regulations; the local health authorities should be consulted about these. In principle, all dairy products to be purchased should be processed from pasteurized milk and cream.

Properly dried milk is safe unless it is recontaminated following the drying process. Dried milk should be used in food preparation in the powder form whenever this is feasible. *Dried milk and milk reconstituted from the powder must be protected from contamination by keeping the products covered and by not contaminating them with soiled dippers and ladles.* Reconstituted milk, because of its moisture content, is as hazardous as fresh milk, or even more so, because the milk powder may have become contaminated in storage; the containers and utensils used in reconstituting the milk are not apt to be sterile; and the reconstituted product is apt to be held out of refrigeration for hours by uninformed or negligent kitchen employees. Much education of kitchen personnel needs to be undertaken to make certain that reconstituted milk is treated with as great respect as fresh milk, that is, protected from contamination and kept under constant refrigeration.

Most state health departments do not permit reconstituted milk to be served as a beverage. Consult your local health authority.

Storage   Information on type and length of storage of the various dairy products, reasons for their limited storage, possible hazards due to prolonged storage, signs of quality loss, and what to do about suspect dairy foods is assembled in Table 3.

*Food Service Sanitation Manual, U.S. Dept. of Health, Education and Welfare, Public Health Service. (See Readings for Part Three.)

Table 3. Safety of Milk and Other Dairy Products

Food	Potential Hazard When Purchased	Suggested Type and Length of Storage	Reason for Limited Storage	Possible Hazards from Handling and Prolonged Storage	Signs of Quality Loss (Spoilage)	What to Do About Suspect Foods
Fluid milk (Fresh)	None, if pasteurized, and carton or bottle undamaged.	Refrigerate at 36 to 40 F (2.2 to 4.4 C) in original container; homogenized milk can be frozen. Daily delivery recommended; store no longer than 1 week beyond expiration date on container.	Properly pasteurized and packaged milk has good keeping quality for several days. Once container is opened, use milk within 1 day.	Opening of carton or other container makes milk vulnerable to contamination. Especially vulnerable is milk poured into pitchers, and then (as leftover) returned to container. Milk contaminated with staph may not appear spoiled.	At refrigerator temperature, spoilage by souring is rather rare; psychrophilic bacteria cause bitterness, proteolysis.	Discard.
Buttermilk, sour cream, other fermented milks	None, if pasteurized, and carton is undamaged.	Refrigerate at 36 to 40 F (2.2 to 4.4 C) in original container. Up to 2 weeks.	Good keeping quality, due to products' acidity.	If accessible to air, mold may grow.	Off-flavor; separation; mold growth.	Discard.

(continued)

Table 3. (Continued)

Food	Potential Hazard When Purchased	Suggested Type and Length of Storage	Reason for Limited Storage	Possible Hazards from Handling and Prolonged Storage	Signs of Quality Loss (Spoilage)	What to Do About Suspect Foods
Cream (Fresh)	None, if pasteurized and container undamaged.	Refrigerate at 36 to 40 F (2.2 to 4.4 C) in original container. Store no longer than 1 week beyond expiration date on container.	Once container is opened, use within 1-2 days.	Opening of container makes cream vulnerable to contamination. Cream contaminated with staph may not appear spoiled.	At refrigerator temperature, spoilage by souring is rare; psychrophilic bacteria cause bitterness, sliminess.	Discard.
Evaporated milk (Unopened)	None, if can appears normal and has no defects.	Cool (50 F; 10 C) storage preferred, especially if in large supply; after long storage, thickening may take place. Average storage is 6 to 12 months; quicker turnover recommended. Shake or turn cans periodically to prevent thickening.	In very warm (120 F; 49 C) storage thermophilic bacteria may spoil contents.	No health hazards except if leakage occurs (bacteria can enter damaged cans).	Swelling of can; off-odor; putrefaction; discoloration; thickening.	Discard.

Evaporated milk (Opened)		Refrigerate at 36 to 40 F (2.2 to 4.4 C) in original container. Up to 3 days.	Once can is opened, contents are vulnerable to contamination.	Handling may introduce staph and other dangerous microorganisms from hands and soiled can openers.	Off-flavor (however, poisoning bacteria may not noticeably change contents).	Discard.
Condensed sweetened milk (Unopened)	None, if appears normal and has no defects.	In cool (50 F; 10 C) storage, several months.	With long, warm storage, can may corrode and milk may spoil. Generally good keeping quality due to high sugar content except if yeasts take over.	Because of high sugar content, yeasts are the main cause of spoilage; yeasts are harmless.	Leaky cans may have contents of yeasty odor; presence of gas; mold growth.	Discard.
Condensed sweetened milk (Opened)		Refrigerate; store up to 1 week.	Yeasts may eventually cause fermentation.	Few, because of high sugar content. Molds may not be harmless.	Yeasty odor and flavor; bubbly, due to yeast fermentation. Mold.	Discard.
Dry skim milk	None, if Grade A Milk is used, and if there is no later contamination; processing under stringent sanitary control is essential to preclude hazard.	Tightly closed in cool, dry, preferably refrigerated storage. 3 months; if refrigerated 10 months.	When held in moist storage, there may be caking, and other quality deterioration (kitchens are warm, steamy places).	Contamination.	Stale odor, flavor; mold may be present.	If off-flavor is objectionable, discard. If moldy, discard.

(continued)

Table 3 (Continued)

Food	Potential Hazard When Purchased	Suggested Type and Length of Storage	Reason for Limited Storage	Possible Hazards from Handling and Prolonged Storage	Signs of Quality Loss (Spoilage)	What to Do About Suspect Foods
Reconstituted		No storage preferred; refrigerate at 36 to 40 F (2.2 to 4.4 C) for no more than 1 day. Do not hold at kitchen temperature.	Chance for contamination from environment, utensils, persons handling milk. Use dry milk in powder form whenever feasible.	Contaminants may be those capable of causing foodborne illnesses.	Same as for fluid fresh milk.	Discard, if not used within safe storage time.
Cheese 1. Cheddar type	Cheeses made from unpasteurized milk; check with local health authority about state regulations.	Refrigerate at 36 to 40 F (2.2 to 4.4 C) in original container. Variable storage life (2 weeks to several months). Watch for appearance of mold spots.	Molds may take over soon after cheese is purchased.	No danger from molds used in the processing of certain cheeses. Safety of cheeses contaminated by spoilage molds is uncertain; more needs to be known about danger from spoilage molds.	Mold growth; if surface moist, yeasts may grow and make conditions favorable for bacteria.	If cheese is very moldly (spoilage molds) discard; do not take chances.

| 2. Soft cheeses | Same as cheddar type. | Refrigerate at 36 to 40 F (2.2 to 4.4 C) in original container. Variable storage life (3 days to 1 week; spreads, once opened, 1 to 2 weeks). Cottage cheese 3 days; longer, if preservatives were added. | High in moisture; bacteria can grow in soft, moist cheeses. Keep covered. | If contaminated with staph through unsanitary handling, a hazard may exist without signs of spoilage. | Sliminess; gas production; off-odor; mold growth. | Discard. |

Eggs

Shell Eggs   Purchase graded, sound-shelled eggs. The shell of eggs becomes contaminated with bacteria, including salmonellae, during the process of laying and thereafter. If the eggs are washed, using effective sanitary procedures, the majority of bacteria are removed. Dirty eggs are heavily contaminated.

Inside, high-quality and undamaged shell eggs are practically free from salmonellae. Unclean, cracked, and leaky eggs are dangerous. *Do not use them, ever.* In 1963 the Surgeon General of the U.S. Public Health Service issued the following statement:

> There is sufficient epidemiologic and bacteriological evidence to suggest that everyone should avoid buying and using cracked or unclean eggs. Persons who are ill, especially infants, the elderly, and individuals suffering from gastrointestinal diseases or malignancy should not be fed raw or undercooked eggs. An undercooked egg is one in which the white is not firm . . .

Shell eggs are graded for size and eating quality on the basis of federal or state standards. Almost every state has legislation for egg standards, and these are fashioned after the federal models. Eggs that are graded by approved personnel and meet federal standards may carry the shield depicted in Figure 21(a).

Purchasing and Storing of Shell Eggs   When purchasing eggs, only eggs graded under federal-state supervision should be used. Only eggs with uncracked shells can qualify for U.S.D.A. grades. The purchaser may get assistance in writing specifications by writing to the USDA*. Eggs should be kept under constant refrigeration. To test eggs for quality, break one or several specimens into a plate. Signs of freshness are a firm white clinging to the yolk and a firm yolk that does not break easily.

Processed Eggs   Frozen eggs and dried eggs are used extensively in quantity food service. The broken-out eggs must be pasteurized to kill salmonellae before they are processed further by freezing or drying since these pathogens may contaminate the contents of the egg when the shell is broken. Eggs that were processed under government supervision carry a stamp (see Figure 21(b)).

*Food Safety and Quality Division, U.S. Dept. of Agriculture, Washington, D.C. 20250.

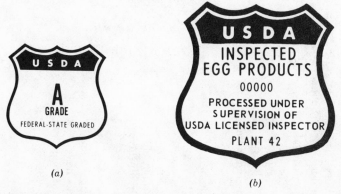

Figure 21. Federal inspection stamp for eggs. (a) Shell eggs. (b) Processed eggs. (Courtesy of U.S. Department of Agriculture.)

Frozen Eggs   Store frozen eggs at 0 F (− 18 C). Defrost in a refrigerator and use promptly. Prevent contamination of the egg mass by keeping cans covered and using clean, sanitized dippers. *Keep under refrigeration at all times.*

Dried Eggs   Keep powder in cool, dry storage—preferably, in refrigerated dry storage. Keep cans tightly closed to keep out moisture and contaminants.

Liquid egg reconstituted from powder is highly perishable and should be used at once. Reconstituted egg should not be contaminated with hands and soiled dippers or through exposure to air, and should not be exposed to lukewarm temperatures for any length of time. *Keep refrigerated.*

From a sanitation as well as managerial point of view, it is advisable to use dried egg in powder form whenever possible rather than reconstituting it.

Storage   Information on type and length of storage of fresh and processed eggs, reasons for their limited storage, possible hazards due to prolonged storage, signs of quality loss, and what to do about suspect items is assembled in Table 4.

Meat and Poultry

Meat   Meat should come from healthy animals, slaughtered and processed under sanitary conditions.

Meat from diseased animals is likely to be contaminated with the

Table 4. Safety of Eggs

Food	Potential Hazard When Purchased	Suggested Type and Length of Storage	Reason for Limited Storage	Possible Hazards From Handling and Prolonged Storage	Signs of Quality Loss (Spoilage)	What to Do About Suspect Foods
Shell eggs	Slight, when eggs have clean, sound shells and are grade A or higher. Cracked eggs, especially leakers, are hazardous and may contain salmonellae; never buy dirty or cracked eggs and never use them.	Refrigeration at 36 to 40 F (2.2 to 4.4 C); preferably 1 week, not longer than 2 weeks.	Quality loss; absorption of odors from other foods in the refrigerator.	Penetration of bacteria through shell.	Nonmicrobial: off-odor, off flavor, thin whites, weak yolk and membranes.	If in doubt, discard.
					Microbial (if bacteria penetrated shell): cloudiness, off-color, off-odor, off-flavor.	Discard.
Eggs broken out of shell		Refrigeration at 36 to 40 F (2.2 to 4.4 C); use tightly covered container; strict sanitation essential.	Excellent medium for bacterial growth.	Contamination from shell of egg and from hands. Danger may exist without spoilage signs.	Whites runny; yolks have loose shape; or no signs.	Discard.
		1 day, no more.				

Frozen egg	Recontamination after pasteurization. Buy from a reliable source.	Store at 0 F (−18 C) or below, up to 6 months. Defrost in refrigerator or in cold running water—never warm. Keep under refrigeration. Use within 1 to 2 days.	Once defrosted, excellent medium for bacterial growth.	Prolonged freezer storage is no sanitary hazard. Once defrosted, danger of contamination from hands, dippers and ladles; danger of bacterial growth due to long holding in refrigeration or holding eggs in warm places during preparation hours.	There may be no obvious changes.	Discard.
Dried egg	Recontamination after pasteurization. Buy pasteurized eggs from a reliable source.	Store in a 40-45 F (4.4-7.2 C) refrigerator in tightly closed containers up to 6 months. Contents of containers, once opened, should be used before they get lumpy.	Powder in opened containers gradually absorbs moisture.	In opened containers: moldiness.	Mustiness, lumpiness, presence of mold.	Discard.

(continued)

Table 4 (Continued)

Food	Potential Hazard When Purchased	Suggested Type and Length of Storage	Reason for Limited Storage	Possible Hazards From Handling and Prolonged Storage	Signs of Quality Loss (Spoilage)	What to Do About Suspect Foods
Egg reconstituted from powder		Use at once. If storage is absolutely necessary, 1 day storage under strict refrigeration, 36 to 40 F (2.2 to 4.4 C).	Excellent medium for bacterial growth. Contaminants of reconstituted egg may stem from powder itself, unsanitary equipment, the human body, dust, and other sources.	Growth of food poisoning salmonellae, staphylococci, and other accidental contaminants.	May not show any signs.	Discard.

animals' pathogens, which may include food-poisoning bacteria such as salmonellae and *Clostridium perfringens*. Under government inspection, diseased animals are eliminated before slaughter and all carcasses are again inspected for signs of disease after slaughter. The killing, cutting, and packing procedures are scrutinized for sanitation. Premises are inspected for clean maintenance, and personnel for freedom from communicable diseases.

The seal of inspection on meat assures that it comes from healthy animals that were slaughtered and processed under sanitary conditions. Inspection may be done either by federal or state agencies if sold within the state. If shipped across state lines, the meat must be inspected for wholesomeness by the U.S. Department of Agriculture. Meat inspected and passed by federal agents carries the round marks depicted in Figure 22 (*a*) and (*b*); it should be present on every inspected carcass, but does not appear on every cut.

At delivery, the cuts should be carefully checked for signs of quality in regard to color, odor, packaging, and temperature. There should be no discoloration; odor should be fresh, not sour; texture elastic, and packaging undamaged and clean. Fresh meat should be cold, with frozen items solidly frozen.

Poultry Like meat, poultry must be inspected for wholesomeness. This must be performed by federal agents when the poultry is to be shipped in interstate trade. [See inspection stamp, Figure 22(*c*)] The sanitation of the processing operations and the plants themselves are also judged. The federal agency in control of wholesomeness is the USDA. However, states may also be allowed to inspect if their standards meet those of the federal government. In addition, there is the possibility for grading poultry for quality, on a voluntary basis; grading is done cooperatively between the USDA and the processor.

Since poultry is highly perishable, it should be closely inspected at delivery for good color, fresh odor, firm texture, and lack of sliminiess.

Does inspection for wholesomeness guarantee complete freedom from bacteria?

The inspection service cannot make such guarantees. Even healthy animals harbor in their intestinal tract bacteria capable of causing foodborne illnesses. Also, it must be remembered that, like humans, animals can become carriers of disease-producing germs such as salmonellae after they have recovered from the disease. Carriers look healthy. When meat animals and poultry are killed and dressed, the intestinal contents containing pathogenic bacteria may accidentally contaminate the meat. Other sources

Figure 22. Federal inspection stamp for meat and poultry. (a) Fresh and cured meat. (b) Canned and packaged meat products. (c) Poultry. (Courtesy of U.S. Department of Agriculture.)

of contaminants are the employees who cut, process, and package the meat. Even heatlhy persons can harbor food poisoning bacteria in and on their bodies.

Which products are most likely to be contaminated?

The extent of contamination of meat and poultry with undesirable bacteria depends on many factors: plant sanitation, working skill and habits of workers, and personal hygiene of workers—to name a few.

Contamination with bacteria is usually lowest in muscle meat and highest in ground meat, organ meats and cuts such as feet, tails, necks, and wings. Muscle meat can become contaminated in the inside when a cut is deboned and rolled. This is true for cuts like rolled rib roast as well as for turkey rolls or deboned hams. As we will see, this fact is important to remember when preparing menu items that are meat or made with meat (Part Three, Sections F and G).

Storage of Meat and Poultry

Meat and poultry should be refrigerated separately from other foods. Information on type and length of storage of meat and meat products, and poultry and poultry products is presented in Tables 5 and 6. Given in the

Table 5. Safety of Meat

Food	Potential Hazard When Purchased	Suggested Type and Length of Storage	Reason for Limited Storage	Possible Hazards From Handling and Prolonged Storage	Signs of Quality Loss (Spoilage)	What to Do About Suspect Foods
Fresh meats (Roasts, steaks, and chops)	Presence of salmonellae, C. perfringens, fecal streptococci, Staphylococcus aureus.	Refrigerate at 30 to 36 F (−1.1 to 2.2 C). Wrap meat loosely. Loosely wrapped: 3 to 5 days or less. Tightly wrapped: 2 days or less. Never keep meat at room temperature except during actual preparation.	With loose wrapping, meat surfaces are dry, retarding bacterial growth; tight wrapping causes sliminess.	Growing bacteria may not change meat noticeably even when present in high numbers.	Discoloration to brown and gray; off-odor, slight to putrid; mold growth appears late in spoilage process.	If in doubt, discard. Spoiled meat may be dangerous. Growth of salmonellae and staph not noticeable.
Ground and stew meat	Same as steaks, roasts, and chops, except that higher bacterial count must be ex-	Refrigerate at 30 to 36 F (−1.1 to 2.2 C). Freeze, unless designated for immediate use. 1 to 2 days,	Although high original quality is very important for keeping quality, all ground and-cut-up	Ground and stew meats subjected to much handling, especially ground meats; bacteria	Same as roasts, steaks, and chops.	Same as steaks, roasts, and chops.

(continued)

Table 5. (Continued)

Food	Potential Hazard When Purchased	Suggested Type and Length of Storage	Reason for Limited Storage	Possible Hazards From Handling and Prolonged Storage	Signs of Quality Loss (Spoilage)	What to Do About Suspect Foods
	pected, especially in ground meat.	preferably only 1 day. Longer storage possible if original quality high and refrigerator temperature is low (near 32 F (0 C).	meats have an excellent chance to spoil fast.	are spread throughout; grinding releases juices from meat cells; oxygen penetrates interior of meat—all these factors favor bacterial growth and there is much surface for bacteria to grow on.		
Variety meats	Same as roasts, steaks, and chops.	Refrigerate at 30 to 36 F (−1.1 to 2.2 C). Loosely wrapped: 1 to 2 days.	Highly perishable.	Same as steaks, roasts, and chops.	Same as roasts, steaks, and chops.	Discard.

Hams (Tenderized, not fully cooked)	Heat-processed, but not sterile; occasionally fecal streptococci and staphylococci (resistant to heat and salt) survive the processing.	Keep refrigerated at 30 to 36 F (−1.1 to 2.2 C); 1 week only for best quality; 1 to 3 weeks is safe, unless mishandled (left out of refrigerator).	Not sterile; highly perishable.	With prolonged storage, contaminants may build up to high numbers and staphylococci may fabricate heat-resistant toxins. Signs of these activities not noticeable.	Discoloration; off-odor; softening; mold growth, if air is available. Iridescence on sliced ham does not indicate spoilage.	If in doubt, discard. Beware of mold growth.
Hams (Tenderized, "fully cooked"; mildly cured, sometimes smoked; internal temperature of 150 F (66C) reached)	Hams are heat-processed but not pressure-cooked; occasionally heat-resistant bacteria survive. Although the hams may come temperature packaged and in cans they are not sterile in the sense that pressure-cooked, canned hams are.	Refrigerate at all times (including at vendor's premises). Storage times same as above.	Not sterile; highly perishable.	With prolonged storage, contaminants may build up to high numbers.	Discoloration; off-odor; softening; mold growth if air is available.	If in doubt, discard. Beware of mold growth.

(continued)

Table 5. (Continued)

Food	Potential Hazard When Purchased	Suggested Type and Length of Storage	Reason for Limited Storage	Possible Hazards From Handling and Prolonged Storage	Signs of Quality Loss (Spoilage)	What to Do About Suspect Foods
Bacon, cured	Fecal streptococci contamination. Sliced bacon is handled more than slab and may receive staph during handling.	Refrigerate at 30 to 36 F (−1.1 to 2.2 C) no more than 1 to 2 weeks.	Becomes moldy quite readily, especially bacon in opened packages. Bacterial spoilage found in unopened packages of sliced bacon.	Contaminants may build up to high numbers.	Off-odor; sliminess; mold.	If in doubt, discard. Do not overstock on bacon.
Cold cuts (Including frankfurters)	Staphylococci from handling are hazard.	Refrigerate at 30 to 36 F (−1.1 to 2.2 C) wrapped with some access to air. Original quality and moisture content affect keeping quality. High-moisture cuts: 3 to 5 days; franks, approx-	Sanitary hazards and keeping quality are related to sanitation practiced in processing plant and in food-service establishment.	Handling, such as slicing and wrapping, may introduce staphylococci or mold.	Discoloration; off-odor; sliminess; mold.	Discard.

imately a week. Hard sausage: 2 or more weeks; inspect weekly for molds.

| Frozen meats | Same as fresh meats. | Freezer-store at 0 F (−18 C) or below. It is good to practice rapid turnover of frozen foods. The limits for storage are these: Beef, lamb: 9 to 12 months. Beef, ground, up to 3 months. Pork: 6 months. Veal: 6 months. Sausage, ham, slab bacon: 1 to 2 weeks. Beef liver: 3 to 4 months. Pork liver: 1 to 2 months. | Assures high quality. | Prolonged freezer storage does not make frozen meat unsafe, but quality suffers. Safely defrosted, the meat should be used within 24 hours of thawing; meats that have been thawed for an unknown length of time and warmed beyond 45 F (7.2 C) are suspect, even if no spoilage signs are noticed. DO NOT RE-FREEZE. | Meat that accidentally has defrosted and remained thawed for unknown periods above 45 F (7.2 C) is suspect and cannot safely be used. Discard. |

Table 6. Safety of Poultry, Fish, and Shellfish

Food	Potential Hazard When Purchased	Suggested Type and Length of Storage	Reason for Limited Storage	Possible Hazards From Handling and Prolonged Storage	Signs of Quality Loss (Spoilage)	What to Do About Suspect Foods
Poultry (Fresh)	Primarily, *Salmonella* contamination; staph, *C. perfringens* and fecal strepotocci may be other contaminants.	Refrigerate at 30 to 36 F (−1.1 to 2.2 C) loosely wrapped; remove giblets from original wrap, rewrap loosely. 1 to 2 days (freeze, if storage must be extended beyond a few days).	Rapid deterioration, due to bacterial and enzyme activity.	Spoilage signs may not be evident, but salmonellae are a health hazard.	Off-odor; discoloration; sliminess.	Discard if spoiled.
Poultry (Frozen)	Same as fresh poultry.	Freezer, at 0 F (−18 C) or below. Chicken, 6 to 8 months. Turkey, 4 to 5 months. Giblets, 2 to 3 months.	Assures high quality.	Safely defrosted, poultry should be used soon after thawing. DO NOT REFREEZE.	Prolonged freezer storage does not make poultry unsafe, but quality suffers.	Poultry that has accidentally defrosted and remained thawed for unknown periods above 45 F (7.2 C) is suspect and cannot safely be used.

Food						
		Cooked poultry, up to 3 months.				
Fish and Shellfish (Fresh, frozen, or smoked)	If harvested from sewage-polluted waters, human pathogens may be contaminants of these foods. *Clostridium botulinum* Type E occurs in certain waters and fish may become contaminated with it. For both quality and safety, purchase from a reliable vendor.	*Fresh fish:* refrigerate, loosely wrapped, at 30 to 34 F (-1.1 to $+1.1$ C); or pack in ice and refrigerate; or freeze if fish cannot be used within 2 to 3 days. *Fresh shellfish:* refrigerate in original, covered container; 1 to 5 days. *Frozen fish and shellfish:* hold at 0 F (-18 C) or below, 3-4 months (fatty fish no more than 3 months). *Smoked fish:* if to be held longer than 3 days, freeze; 3-4 months.	*Fresh:* highly perishable; deterioration due to bacterial activity, autolysis (self-digestion), rancidity. *Frozen and smoked:* may become rancid. Also, *C. botulinum* Type E can multiply at refrigerator temperatures of 38 F (3.3 C) and above.	*Refrigerated fish:* contaminants may multiply. *Frozen items:* fish and shellfish that have been thawed for unknown time and have warmed to and beyond 38 F (3.3 C) are suspect. DO NOT REFREEZE.	*Fresh fish:* flesh softens, eyes are sunken, flesh has strong or off odor. *Smoked fish, oysters and clams:* abnormal appearance, sour odor, flavor or other off-odor.	Discard.

same tables are the reasons for limited storage of these items; possible hazards because of prolonged storage; signs of quality loss; and what to do about suspect foods.

Facts to remember about meat and poultry:

1. Purchase government-inspected meat and poultry.
2. Store meats and poultry by themselves. Do not store meat long, especially ground meat, organ meats, and other cuts apt to have high bacterial counts.
3. Expect even high-quality meat and poultry to be contaminated with some food-poisoning bacteria. Therefore, use all the precautions in their preparation that are discussed throughout Part Three, Sections F and G.

Fish and Shellfish

Fish and shellfish are highly perishable foods. All fish and shellfish should be bought from reputable vendors. Fresh fish should be packed in ice; frozen fish should be kept solidly frozen.

Fish and shellfish should come from unpolluted water. When raw sewage is directed into water, human intestinal pathogens such as typhoid and other intestinal pathogenic bacteria, and hepatitis virus, will contaminate the wildlife therein.

Special danger exists when seafood harvested from polluted water is eaten raw or lightly heated. Effective supervision of dangerous harvesting areas is practiced in states where the shellfish processing is an important industry. The Food and Drug Administration, U.S. Public Health Service, cooperates with states in this endeavor by certifying shippers. *Only shellfish from certified shippers should be purchased.* A listing of approved shippers, prepared by the FDA, can be obtained from local health agencies.

Storage   Fish and shellfish should be used promptly. If they must be stored, store them briefly, and store at temperatures close to freezing, or freeze them.

Information on type and length of storage, reasons for limited storage, possible hazards from prolonged storage, signs of quality loss, and what to do about suspect items is presented in Table 6.

Canned Food

Purchase and Storage   Damaged, bulging, leaking, and rusty cans may be dangerous and should be refused. One should not do business with a vendor who dares to deliver such foods. Canned foods should be stored in a cool, dry place. Excessive heat will spoil the contents, and freezing may break the seal of cans and allow bacteria to enter. Among these bacteria may be types injurious to health. Rapid turnover of canned foods is advised, since the quality of all canned foods can only go down, never up.

Information on length of storage, reason for limited storage, possible hazards due to handling and prolonged storage, signs of quality loss, and what to do about suspect items is given in Table 7.

Frozen Precooked Foods

Purchase and Storage Frozen precooked foods have been cooked in processing plants. If they have been prepared in a sanitary processing plant by sanitary methods of handling and if they have been frozen rapidly, and kept frozen solidly up to the time they are used, all should be well. We see that there are many "ifs." *Defrosting, whether it occurs in the vendor's storage, in transit or elsewhere, is of real danger since bacteria may resume growth once the food is thawed.*

1. Purchase frozen menu items from reliable vendors. Even products processed under *good* sanitary conditions may lose their quality and wholesomeness when they are allowed to defrost during transit or storage.
2. Do not overstock. Frozen precooked foods lose quality fast.
3. Upon delivery, move frozen items immediately into freezers maintained at 0 F (−18 C) or below.

Frozen foods should remain solidly frozen at all times; this means at 0 F (−18 C) or below. Information on type and length of storage, reasons for limited storage, possible hazards due to prolonged storage, signs of quality loss, and what to do about suspect food is presented in Table 7.

GENERAL RULES FOR PURCHASING SOUND FOOD SUPPLY

Purchaser

1. Inform yourself of health regulations that apply to the foods you buy and serve.
2. Buy only from reputable vendors. It is advisable to visit the vendors' business premises and check (a) the sanitation of the facility and the food supply; (b) the handling procedures and storage conditions; (c) the adequacy and sanitary condition of refrigerators and freezers; and (d) the sanitary conditions of the trucks.
3. Assure the wholesomeness of potentially hazardous foods.

Table 7. Safety of Canned Foods and Precooked Frozen Foods

Food	Potential Hazard When Purchased	Suggested Type and Length of Storage	Reason for Limited Storage	Possible Hazards From Handling and Prolonged Storage	Signs of Quality Loss (Spoilage)	What to Do About Suspect Foods
Canned foods	None, if properly processed, and if cans are intact. Underprocessed, home canned, nonacid foods may contain *Clostridium botulinum* toxin.	Cool (50–60 F) (10 to 15.6 C), dry storage. Freezing may break seal of can or jar or crack jar; leaks will allow bacteria to enter. Unless cans or jars are damaged, freezing will not make food unsafe. Maximum storage time, 1 year, but 6 months is preferable.	General decrease in quality, especially of low-acid foods. Storage at very warm temperatures allows growth of thermophilic bacteria.	Reentry of bacteria in jars and cans, due to disintegration of seals.	Color change (red foods may lighten; colored foods fade in plain tin cans). All canned goods are gradually reduced in quality if held above 50 F (10 C).	Do not taste foods in cans that leak; that are rusty, corroded or have a broken seal; or whose contents spurt out or are bubbly, off-color, or off-odor.

Food					
Frozen precooked foods	Pathogens introduced from hands and equipment; delayed freezing; warm-up during thawing.	Freezer at 0 F (−18 C) or below. Length of storage depends on product. Do not store anything over 3 months.	Quality disintegrates at different rates in the various frozen precooked foods.	No health hazard from prolonged storage in freezer; danger arises when food thaws and warms beyond 45 F (7.2 C) DO NOT REFREEZE.	Dehydration; discoloration; large ice crystals.	Potentially hazardous foods that thaw and warm to 45 F (7.2 C) are suspect. Discard.

Dairy Products   Milk and cream should be pasteurized, and other dairy products should be processed from pasteurized milk and cream.

Eggs   Purchase graded, sound-shelled eggs. Frozen and dried eggs have been pasteurized before they were frozen or dried.

Meat and Poultry   Purchase government-inspected meat and poultry.

Fish and Shellfish   Buy oysters, clams, and mussels from sources approved by the state shellfish authority or certified by the state or origin. Fish purchased locally should come from vendors with a reputation for handling fresh, wholesome fish.

Canned Foods, Staples   Buy canned foods from reliable vendors. Beware of bargain foods, unless you have convinced yourself that they deserve to be called bargains. Cheap foods are often passed on to the customer because they have been damaged due to one cause or another. Damaged cans will become unsafe when microorganisms enter; staple items in sacks or cartons are in danger of being damaged through the activity of rodents and insects.

Precooked Frozen Foods   Purchase frozen foods from reputable vendors with facilities to keep frozen items in the solidly frozen state while in storage and transit. Beware of bargain goods.

GENERAL RULES FOR CHECKING AND STORAGE OF DELIVERIES

1. Personnel assigned to the task of receiving the deliveries should have the necessary storage space ready when the goods arrive.
2. Deliveries should be checked against the specifications. Perishables should be handled first. To ascertain the temperatures of chilled and frozen items, a correct thermometer and appropriate methodology of taking measurements should be used. (See pages 189–191).
3. Deliveries should be inspected for their condition and for signs of spoilage, damage, insect infestation, and general mishandling. Spoiled, damaged, and mishandled foods should be refused.
4. Deliveries that do not go to the kitchen for immediate use should be stored at once in appropriate storage area.
5. Staples are stored in "dry storage." This area should be:
   (a) Above grade level to protect it from danger of flooding.
   (b) Free from overhead water or sewer pipes to guard against danger of dripping water or sewage.
   (c) Free from floor drains to guard against danger from backup of waste materials, including sewage.

(d) Cool, dark, well ventilated, and lighted.

(e) Well maintained—neat and orderly.

(f) Free from insects and rodents.

(g) Provided with platforms, shelves, racks, bins, or dollies that raise the food above floor level to guard against contamination and danger from flooding; and for easy cleaning of the floor.

(h) Sufficiently roomy to avoid crowding of food and food contact with the wall.

(i) Provided with brooms and waste receptacles.

6. Perishables are stored under refrigeration, either in chilled storage (refrigerators) or in freezers. Some important requirements are:

(a) Separate refrigerators should be provided for dairy products and eggs; meat, poultry, and fish; fruits and vegetables; and cooked foods.

(b) Cold storage should be sufficiently ample to accommodate all food at peak periods such as long weekends and holidays; accommodate frozen items to be defrosted (frozen eggs, roasts, turkeys, hams, etc.); and answer the heavy demands of summer weather.

(c) There should be sufficient shelving to assure that even at peak periods the refrigerated materials may be spread out for good air circulation and rapid cooling.

(d) Cold storage should be placed conveniently. If this is not feasible for long-time storage, additional refrigerators should be provided near the area of food preparation. These refrigerators would be used for items used within the working day. *Refrigeration that is not easily accessible will not be used by a cook or baker.*

(e) Thermometers should be installed in every refrigerator and checked frequently and regularly.

(f) Metal and/or plastic containers should be provided for storing food items in a sanitary manner. Storing food in cartons, wooden crates, bushel baskets, and the like is an undesirable, unsanitary practice.

(g) Food should not be placed directly on shelves plated with cadmium; this metal may cause food poisoning.

(h) Food should not be crowded.

(i) Open wire shelves rather than solid shelves should be provided; and shelves must not be lined with paper, cardboard, and the like because these products hinder circulation of air.

Responsibilities of management are to:

(a) Provide sufficient, adequate storage for chilled and frozen foods.

(b) Provide refrigeration and shelving that permit good air circulation.

(c) Keep refrigerators and freezers in good working order.

(d) Provide directions for cold storage of various foods to the employees.
(e) Have refrigerators and freezers routinely cleaned and defrosted.
(f) Check coolers and freezers for orderly arrangement, cleanliness, and lack of crowding.
(g) Check for cleanliness of the surrounding area.
(h) See to it that food is rotated; the new supply should go to the back and previously stored items to the front so that they will be used first.

Responsibilities of employees are to:

(a) Store the various foods in the space designated for them.
(b) Store food according to the directions provided by management.
(c) Avoid crowding the refrigerators with food; air circulation is necessary for rapid cooling.
(d) Not clutter refrigerators and freezers with unwanted items that are eventually discarded anyhow.
(e) Keep refrigerators orderly and neat, wipe up spills, and so on.
(f) Not place trays directly on other foods; tray bottoms may contaminate food, and, air circulation is hampered.
(g) Follow a schedule and directions, provided by management, to clean and defrost refrigerators regularly.

SECTION E

Sanitation-Conscious Food Handlers: Their Health, Personal Hygiene, Work Habits, Interest, and Attitude

The health of the food handler plays an important part in sanitation. Food handlers can be sources of bacteria causing illness in others through transmission of pathogens causing disease or by contaminating food with microorganisms capable of causing foodborne intoxications and infections.

Experience has shown that a great number of foodborne illnesses come from workers who were ill, practiced poor personal hygiene, or applied unsanitary food-preparation methods. When food poisoning or food infection strikes us down, our food may have been contaminated through the food handler's sneeze, or through pus from infected wounds or pimples, or through fecal matter clinging to the hands.

Even the healthy person may harbor food-poisoning bacteria in the nose, mouth, throat, and intestinal tract. Therefore, the personal hygiene of a food handler is of great sanitary importance. Personal hygiene refers to the cleanliness of a person's body.

A food handler's hand habits and working habits are important in connection with sanitation. Undesirable hand habits during food preparation are scratching one's head, arranging one's hair, or touching one's nose or mouth. Thereby, bacteria from the person's body are transferred to food and sow the seed of food poisoning. Also, a person's clean or sloppy working habits can make the difference in reducing or increasing the chances of spreading bacteria in the kitchen.

A food handler's interest in sanitation and cooperative attitude are of extreme importance in practicing sanitation. To be interested, the individual must first learn about sanitation; he or she must understand what it is, what lack of sanitation does, and what good sanitation can do. And he or she must understand, and be impressed by, the fact that *the food handler plays an important part in the practice of sanitation. When applying the good habits learned, the food handler must become convinced of the fact that the supervisor sees and appreciates this.*

HEALTH AND PERSONAL HYGIENE OF EMPLOYEES

Health and sanitary personal habits and a knowledge of safe food practices are most important qualifications in a food handler. Therefore, management has to concern itself with:

Hiring people who are in good health.
Training food handlers in acceptably clean personal habits.
Training food handlers in sanitary techniques of food handling.
Constant and careful supervision.

Management and employees should cooperate in the effort to keep ill employees from working with food and with equipment and utensils used in preparing and serving food. Examples of important human illnesses that may be transmitted through food are diseases of the respiratory tract such as the common cold, sore throat, scarlet fever, tonsillitis, pneumonia, sinus infections, and tuberculosis (Table I, Appendix). Other diseases that may be transmitted are intestinal disorders such as typhoid fever, dysentery, Asiatic cholera, and infectious hepatitis (Table II, Appendix). In some of these diseases the germs may remain with the person after he or she has recovered; the person may be a carrier without knowing it.

In the Food Service Sanitation Manual of the Public Health Service,* the following requirment regarding the health of food service personnel is given:

No person, while infected with a disease in a communicable form that can be transmitted by foods or who is a carrier of organisms that cause such a disease or while afflicted with a boil, an infected wound, or an acute respiratory infection, shall work in a foodservice establishment in any capacity in which there is a likelihood of such person contaminating food or food-contact surfaces with pathogenic organisms or transmitting disease to other persons.

This regulation is stringent indeed and may serve as model. It would be difficult to carry it out exactly to the letter of the law. But it should be heeded as closely as is feasible. For example, a person recovering from a cold might be permitted to return to work provided he or she does not handle food or utensils used in food preparation and service. The manager or supervisor should make it clear to the employee that it is not punishment to be sent home or temporarily shifted to another—perhaps more menial—job. The employee should understand that this is for the good of the "cause," which is sanitary foodservice.

*U.S. Dept. of Health, Education and Welfare, Public Health Service, Food and Drug Admin. *Food Service Sanitation Manual.* (See Readings for Part Three).

Responsibilities of management are to:

1. Require a new employee before the person starts working to have a physical examination to show whether he or she has an acute communicable illness or is a carrier of a communicable disease.
2. Keep a watchful eye on employees for signs of illness that might be of a contagious nature; and for the presence of infections such as infected burns, sores, and boils.
3. Require a physical examination of an employee after the person has been ill with a communicable disease and before he or she resumes work to assure that he or she has not become a carrier.
4. Stimulate employees to become health-conscious; educate them to hear about and understand certain basic facts about health; encourage employees to report to their supervisors any respiratory infections and gastrointestinal upsets they may have. Here, an ounce of prevention is worth a pound of cure. In many ailments, the first stages of illness are the very infectious ones; also, prompt care and bed rest usually favor faster recuperation of the patient and a quick return to work.
5. Provide the facilities that are essential for the employees to keep themselves clean.
   (a) Provide dressing rooms equipped with lockers, sinks, soap, and paper towels or an air dryer. Have the rooms cleaned regularly; inspect them regularly.
   (b) Provide uniforms, laundering services, or both.
   (c) Provide hand-washing sinks in the kitchen, hot water (preferably with knee-controlled or foot-controlled valves), soap, and paper towels or air dryers.
6. Apply diligent supervision.
7. Develop a training program in sanitation; train all new employees; and continue to train all employees, new and old. (See Part Four.)

Responsibilities of employees are to:

1. Keep themselves in good health. A healthy person feels energetic, sleeps well, performs work without undue fatigue, has lustrous hair, clear skin, is free from respiratory and gastrointestinal disorders and other physical ailments.
2. Be health-conscious; learn about certain health facts; learn how their own health may affect that of others; they will be surprised and pleased to see how important their knowledge about health, and their attitudes about practicing good health habits, can be to the welfare of others.

3. Report to their supervisor when they have been injured, including minor injuries such as cuts and burns. Report any skin eruptions, boils, and the like.

4. Report to their supervisor any abnormal condition of their respiratory system (head cold, sinus infection, sore throat, tonsillitis, bronchial and lung disorders) and intestinal disorders, such as diarrhea.

5. Report to their supervisor if the supply of soap or towels in washrooms is low or gone.

6. Practice personal cleanliness.
    (a) Take a bath or shower daily.
    (b) Use a deodorant.
    (c) Wash hair at least once a week and keep it neat.
    (d) Keep nails clean, neat, and well-trimmed.
    (e) Wear clean undergarments daily.

7. Prepare themselves for work in a systematic fashion.
    (a) Brush and comb their hair.
    (b) Wear a hairnet (women) or cap (men).
    (c) Get into clean, comfortable shoes.
    (d) Change into their uniform; hang street clothes in the locker; do not hang clothes in the toilet or kitchen.
    (e) Clean hands and fingernails thoroughly. Wash hands with plenty of hot water and soap and dry hands on disposable towels or before an air dryer.

8. When at work, control bad hand habits. Do not scratch head or other body parts, arrange hair, stroke a moustache, pick pimples, and so forth. If employees accidentally do these things, they should wash their hands afterwards.

9. Avoid coughing or sneezing.

10. Wash their hands frequently and every time after having performed one of these activities:
    (a) Visited the toilet.
    (b) Coughed or sneezed into hands or handkerchief.
    (c) Smoked; the saliva from the butt of the cigaret or cigar will contaminate the hands.
    (d) Handled boxes, crates, packages, and other soiled articles.
    (e) Handled raw meat, poultry, shell eggs, fish, or shellfish.
    (f) Handled garbage.
    (g) Handled money.
    (h) Handled anything soiled.

11. Keep hands and fingers out of food; do not taste food from fingers; use a clean tasting spoon each time.

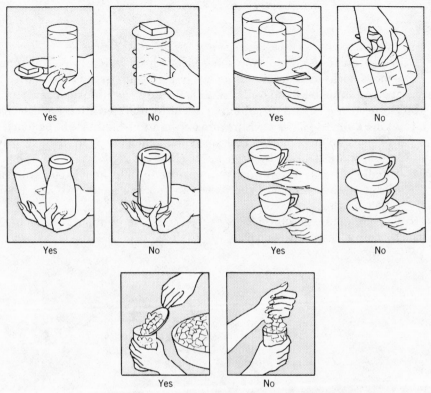

Figure 23. Carrying and serving food in a sanitary manner.

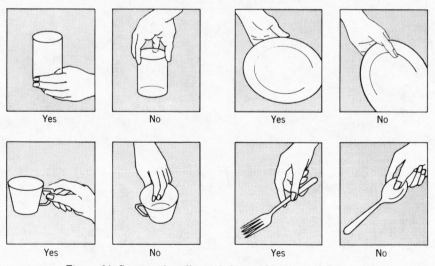

Figure 24. Sanitary handling of china, glasses, and flatware.

12. Use utensils for preparing food as much as is feasible. Use tongs, spoons, forks, etc., for handling ice, butter, rolls, breads, pastries, etc., instead of using hands (Figure 23). Be sure the utensils are clean.
13. Observe "no smoking" rule in areas of food preparation and service.
14. Avoid touching parts of china and eating utensils that will be touched by the customer's mouth. Pick up serving and eating utensils by their handles, and glasses by their bases (Figures 23, 24).
15. Handle plates in a sanitary manner; the thumb should touch the rim of the plate only, never the plate itself or the food on it. Use plates under bowls.
16. Use plastic disposable gloves if food must be manipulated by hand. Dispose of used gloves.

SECTION F

Sanitary Procedures
of Preparing and
Holding Food: General
Principles

The wholesomeness of prepared food can and must be safeguarded through sanitary practices in kitchen and storage areas, which includes effective control of food temperatures at all times. In sanitary foodservice only clean, safe food is served to the public.

We have described the techniques applicable to the cleanliness of the physical plant; and the methods for cleaning and sanitizing equipment, utensils, china, glasses, and flatware. We have shown ways for the purchase of a sound food supply and its sanitary storage. We will now be concerned with the following aspects of sanitary practice:

(a) Sanitary methods of handling food in the kitchen.
(b) Safe holding of prepared food using appropriate temperatures.
(c) Safe cooking procedures.
(d) Safe cooling procedures.

To safeguard the wholesomeness of finished menu items at the time they are served, the following rules need to be observed:

Rule I   Start out with wholesome food material.
Rule II   Remove by thorough washing, soil and chemical residues that must be expected to be present on certain foods.
Rule III   Protect food from contamination with poisonous substances and disease-producing bacteria throughout its preparation, storage, and service.
Rule IV   Thoroughly heat those foods that can be expected to harbor illness-producing microorganisms, whenever this is feasible.
Rule V   Prevent multiplication of microorganisms in the food and on equipment used in its preparation, storage, and service.

Rules I, III, and V are applicable to all menu items. Rule II is particularly important in the case of fruits and vegetables eaten raw. Rule IV can only be applied to items that tolerate thorough heating. For example, a meat stew is always heated to a simmer or boil (which is thorough heating) in routine cooking. Hollandaise would curdle if thoroughly heated; and, of course, salads cannot be heated at all.

PRACTICE RULE I—START OUT WITH WHOLESOME MATERIALS

Follow the suggestions given in Part Three, Section D.

PRACTICE RULE II—WASH FOOD TO REMOVE SOIL

Foods to Wash

Not all foods can or need to be washed. However:

- Wash all fruits and vegetables to be eaten raw.
- Wash fresh vegetables to be cooked.
- Wash dried fruit and raisins.
- Wash raw poultry, fish and variety meats; the body cavity of poultry is especially rich in salmonellae capable of causing food poisoning.

How to Wash

Fruits and Vegetables   Use lukewarm running water. Wash fresh greens thoroughly to remove every bit of dirt. Rinse washed products with cold running water and drain.

In case of suspected insect infestation, soak fresh vegetables in cold salted water for 20 minutes. Salt will cause insects to rise to the top of the water.

Poultry, Fish, and Variety Meats   Wash items shortly before cooking them. Use lukewarm water followed by a cold rinse. Drain well. If items are not to be cooked immediately, keep them chilled until the time for cooking has come.

PRACTICE RULE III—PROTECT FOOD FROM CONTAMINATION

Contamination of food with poisonous substances and certain types of bacteria may cause food poisoning in persons eating such contaminated food.

Contamination From Poisonous Substances Other Than Bacterial Toxins

Most commonly involved are chemicals that get into food by mistake. Among them are cleaning compounds, polishes, insect powders, and the like (see Part Two, Section D).

Protective Measures   All chemicals must be stored separately from food, never in the kitchen, pantry, and other places where food is handled and stored. *There is no excuse for slipshod housekeeping.*

Contamination from Bacteria Capable of Causing Foodborne Illnesses

In order to understand the reasons for prescribing certain protective measures in the handling of food, the material presented in Part Two, Sections B and C should be carefully read. Here are some important facts in brief that underlie procedures devised to protect our food:

- Microorganisms causing communicable (or catching) disease may be passed on from one person to others by way of food. Therefore, *persons ill with such diseases must not be allowed to handle food; the same holds true for carriers.* Carriers are persons who harbor disease-producing bacteria, although they appear healthy.
- Microorganisms capable of causing food poisoning frequently originate from the body of the food handlers. Even healthy people may harbor numerous microorganisms, capable of causing foodborne illness, in and on their bodies. Other sources are soil and dust; sewage-polluted water and food; and vermin such as rodents, flies, and cockroaches. Still others are foods of animal origin such as meat, poultry, and shellfish.
- Food-poisoning bacteria from the above sources will in turn contaminate work surfaces, equipment, and utensils used in the preparation, service, and storage of ingredients and prepared menu items.

PROTECTIVE MEASURES

Management and food handlers must fully cooperate in making sanitary measures work. Sanitary rules must be set up that are sound as well as practical. The employees must be trained in following them and must be supervised in practicing them. Everyone must see his or her own role in the important task of preparing food the sanitary way.

Management has the obligation to:

1. Be informed on the dangers of unsanitary practices in food preparation and service and convinced of the importance of sanitary food handling in the establishment.
2. Provide a clean facility.
3. Provide the sanitary equipment necessary for doing a sanitary job.
4. Provide utensils for tasks that can be successfully used in place of hands, such as forks, tongs, and spoons; and see to it that these are kept clean.
5. Provide plastic gloves for tasks that may require the human hand, such as mixing salads, shaping meat or fish patties, and the like; see to it that gloves are always available so they will not be used twice and that gloves are used for one task only, and discarded after use.
6. Educate the food handler in the fundamentals of sanitation in food service and train the person in the appropriate techniques.
7. Set up directions for employees to follow.
8. Provide supervision.
9. Keep food handler's interest alive through praise for jobs done well and through constructive criticism when the person fails. However, when a food handler, after careful training, consistently fails to use sanitary techniques, he or she is not the person to keep in the food department.

Food handlers have the obligation to:

1. Be willing to be educated in the fundamental facts about sanitation in foodservice on which sanitary practice is based.
2. Understand that they play a very important role in the total sanitary program of the establishment.
3. Understand that they—as healthy people—are likely to carry food poisoning bacteria on and in them and do their best to keep these bacteria out of the food they prepare.
4. Be willing to learn sanitary techniques and use them without fail, all hours of the working day.
5. Be health-conscious, of clean appearance, and practice good personal hygiene; report to the supervisor any signs of an oncoming illness such as sore throat, head cold, flu, nausea and diarrhea, and the presence of infected sores, burns, or other infections on the body.
6. Keep undesirable hand habits under control such as fingering nose, hair, moustache, or other body parts.
7. Refrain from sneezing or coughing onto food; sneeze or cough into a handkerchief or—in case of "surprise attack"—into the hands. *In*

either case they immediately wash their hands to remove dangerous bacteria.

8. Understand that a visit to the toilet will contaminate hands with dangerous bacteria; therefore, they wash hands thoroughly before returning to work.
9. Understand that handling raw meats, poultry, sea food and the shells of eggs may deposit harmful bacteria on hands, and wash hands before touching other foods.
10. Understand that hands need to be washed frequently throughout the working day, and immediately after handling soiled articles.
11. Use clean, sanitary equipment.
12. Keep work surfaces clean and orderly, clean up spills promptly.
13. Keep hands out of food as much as is possible and use appropriate clean utensils instead.
14. Use gloves whenever their use is prescribed by management.
15. Never use a tasting spoon twice.
16. Always touch spoons, knives, forks and other eating utensils by the handles, glasses by their bases, cups by their handles, plates by their rims.
17. In serving never stack dishes in such a way that the areas touching the customer's lips become contaminated.
18. Prevent cross-contamination of bacteria during preparation of food (see remaining items on list).
19. Use, whenever possible, separate cutting boards, blocks, tables, grinders, slicers and other utensils for raw and cooked food.
20. If same equipment and utensils must be used, thoroughly clean and sanitize equipment used for raw food after each use.
21. Thoroughly wash hands after handling raw food.
22. When storing food, arrange it so bottoms of containers (pans, trays, and pots) will not contaminate food stored underneath, since containers that have touched unsanitary surfaces are loaded with bacteria. If one food is stored above another, bottoms of containers must be sanitary.
23. Protect food from dust, flies, cockroaches, and rodents.
24. Never leave food out overnight.
25. Cover food except when it is actually manipulated.

The practice of Rule IV and Rule V involves the application of *strict control of temperature.*

The measuring of temperatures needs to be done properly. It requires:

1. Appropriate thermometers.
2. Proper method of taking measurements.

Measuring Food Products

Dial-type thermometers with a range of 0 to 220 F (−18 to 104 C) are commonly used. Federal guidelines* suggest "metal stem-type numerically scaled indicating thermometers, accurate to ±2 F (1 C), shall be provided and used to assure the attainment and maintenance of proper internal cooking, holding, or refrigeration temperatures of all potentially hazardous foods." This type of thermometer is commonly known as a bimetal thermometer. For refrigerated storage federal guidelines suggest the use of a "numerically sealed indicating thermometer, accurate to ±3° F (±1.5° C), located to measure the air temperature in the warmest part of the facility and located to be easily readable. Recording thermometers, accurate to ±3 F (±1.5 C) may be used in lieu of indicating thermometers." For hot storage these same federal guidelines suggest the use of a "numerically scaled indicating thermometer accurate to ±3 F (±1.5 C), located to measure the air temperature in the coolest part of the facility and located to be easily readable. Recording thermometers accurate to ±3 F (1.5 C), may be used in lieu of indicating thermometers. Where it is impractical to install thermometers on equipment such as bainmaries, steam tables, steam kettles heat lamps, cal-rod units, or insulated food transport carriers, a product thermometer must be available and used to check internal temperature."

Use the following precautions when measuring product temperature:

1. The thermometer must be cleaned and sanitized by washing, rinsing, and a subsequent sanitizing procedure. Use separate thermometers for raw and cooked items.
2. Depth of immersion varies with the type of thermometer. Partial immersion of the stem applies to all bimetal thermometers; the proper depth is indicated by the manufacturer.
3. When taking the temperature in the thermal center (warmest when cooling; coldest when heating), be as accurate as possible by taking several readings to determine the hot or cold spots.
4. Allow sufficient time for the instrument to stabilize; there should be no wavering of the pointer for approximately 15 to 30 seconds.
5. Special techniques are needed for measuring temperature of frozen items. A steel thermometer should be inserted under the top layer of items inside a case or other package, or through a hole punched in an unopened case. Allow time (at least 5 minutes) before recording the temperature.

*U.S. Dept. Health, Education and Welfare, Public Health Service, Food and Drug Administration. *Food Service Sanitation Manual.* 1976 Recommendations. DHEW Publication no. (FDA) 78-2081. 1978.

A pocket guide "Thermometers in Food Service,"* prepared by the National Sanitation Foundation, gives valuable information on the characteristics of thermometers, field checking (calibration) of the instruments, and instructions for measuring product temperatures accurately, using sanitary precautions.

PRACTICE RULE IV—THOROUGHLY HEAT ALL FOODS THAT MAY BE EXPECTED TO HARBOR ILLNESS-PRODUCING MICROORGANISMS, WHENEVER THIS IS FEASIBLE

The following foods may be expected to harbor illness-producing microorganisms:

Group 1.   Raw meat—because contamination with salmonellae and *C. perfringens* must be expected.

Group 2.   Raw poultry—because contamination with salmonellae must be expected.

Group 3.   Cracked shell eggs and recontaminated processed egg—because of salmonellae.

Group 4.   All items, raw or cooked, that have been subjected to manipulation by hand or machinery, because of contamination with staph, salmonellae, and perfringens bacteria; these organisms may stem from the food handler, soiled equipment and utensils; raw meat and poultry; and egg shells.

Temperatures to which items should be heated to kill dangerous microorganisms.

Heating to at least 165 F (74 C) is necessary to kill bacteria that do not form spores, such as staph, salmonellae, and streptococci. Trichinae will also be killed by such heating. However, the spores of perfringens bacteria are resistant to heat and require temperatures higher than 165 F (74 C) for kill.

Menu items that in routine cooking are heated to safe temperatures at site of contamination.

Group 1.   *Meats.*   (a) Items that are boiled, stewed, braised, steamed, broiled, and panbroiled are safe.

(b) Roasts that have not been boned and roasts that were boned but not rolled

*National Sanitation Foundation. *Thermometers in Food Service, a Pocket Guide.* Prepared by the Nat. Sanit. Foundation, 3475 Plymouth Rd., Ann Arbor, Mich. 48105.

are also safe even if heated to the rare stage; the meat is sterile on the inside, and outside bacteria are killed. For rolled roast, especially those cooked to the rare stage, special sanitary precautions must be taken (see Section G, Table 16).

(c) Trichinae are killed in pork roasted to the gray stage (170 F; 77 C).

Caution: Boned, rolled roasts, gravies, and items that are made from ground meat may be unsafe.

Group 2.   *Poultry.* (a) Items that are boiled, stewed, fricasseed, braised, steamed, broiled, and panbroiled are safe.

(b) Roast poultry is safe (except large stuffed turkey).

Group 3.   *Eggs and* (a) Hard cooked eggs are safe.
egg-thickened (b) Souffles are safe.
products. (c) Baked custards are safe.

Group 4.   *Items that have* Some of these items are safe only *if* cooked
undergone much to the "done" stage (hamburger) or are
handling heated to boiling temperatures all the way
(grinding, through (casseroles, meat pies, poultry pies
chopping, or and the like).
mincing).

Caution: Some of these items are frequently underheated and are therefore discussed in Part Three, Section G.

Menu items that in routine cooking cannot be heated to safe temperatures because they would lose appeal or are not heated to safe temperatures for other reasons.

Group 1.   *Meat and gravy.* (a) Boned, rolled roasts heated to the rare and medium rare stage are hazardous. The inside of the roast, likely to become contaminated during deboning, does not get sufficiently hot for bacteria, especially spores, to be killed (Table 16).

(b) Gravies may or may not be safe. The danger exists that gravies may become contaminated with perfringens bacteria from meat juice released from the inside of rolled rare roasts. Boiling the finished gravy will make gravies safe, unless the spores are of the heat-resistant type.

(c) Items made with ground meat are hazardous unless cooked to the gray (done) stage.

(d) Hams may be unsafe unless heated to 165 F (74 C), preferably 170 F (77 C), although the trichinae, if present, would be killed at 150 F (66 C).

Group 2.   *Poultry.*

(a) Large (18 pounds and over) roast turkeys that are stuffed are unsafe because heat penetration is slow into the center of the stuffing. Large turkeys should never be stuffed.

Group 3.   *Eggs and egg-thickened products.*

Lightly heated are:

(a) Soft cooked eggs. Therefore, use sound-shelled eggs.

(b) Scrambled eggs. Use sound-shelled fresh or pasteurized processed egg.

(c) Omelets cooked on top of the stove. Use sound-shelled fresh eggs or pasteurized processed egg.

(d) Stirred custards—use sound-shelled fresh eggs or pasteurized processed egg.

(e) Egg-thickened items that are not really hot when the egg is incorporated (stirred custards, puddings, and cream fillings). Use proper heating procedure (Part Three, Section G).

(f) Meringue pies. Use proper heating procedure (Part Three, Section G).

Group 4.   *Certain items that have undergone much handling.*

(a) Deep-fried items such as croquettes routinely receive light heat treatment.

(b) Pan-fried patties like crab cakes and fish cakes routinely receive light heat treatment.

       (c) Meat pies, poultry pies, and casseroles (freshly made-up or frozen) are frequently underheated by default.

       (d) Creamed items are frequently underheated by default.

       (e) Leftovers to be served hot are frequently underheated by default.

       (f) Cream-filled pastries.

Summary of recommendations on heating foods that may be expected to harbor illness-producing microorganisms:

1. Items that are improved by, or can tolerate, without quality loss, high heat should be heated to at least 165 F to be rendered safe (although spores of perfringens bacteria cannot be expected to be killed at 165 F).

    There is no excuse for underheating by default (carelessness, poor planning, or negligence.)

2. Special heating procedures can be used for certain items to assure that dangerous bacteria are killed in the cooking process. The student is referred to Part Three, Section G.

3. In cases where heating adequately to kill bacteria is not possible, protective measures other than application of "killing heat" must be used to assure that the items are safe when served. If we cannot kill the dangerous bacteria, we must—without fail—prevent them from multiplying. (See Rule V. See also Section G "Techniques for Handling Some Specific Hazardous Menu Items").

PRACTICE RULE V—PREVENT MULTIPLICATION OF MICROORGANISMS

We do not live in a sterile world. Bacteria are everywhere. Even if we use care and try to prevent contamination of the menu items we serve, some bacteria will probably find their way into the food and contaminate it. *We must prevent the multiplication of contaminants at every step of food preparation through appropriate time-temperature control.*

1. During actual manipulation in preparation.
2. During hot-holding of prepared food.
3. During chilling and freezing of prepared food.
4. In handling of chilled and frozen food.

Temperature at which bacteria multiply or merely survive, or die, are given in Figure 25.

Bacteria capable of causing foodborne illnesses multiply over a wide

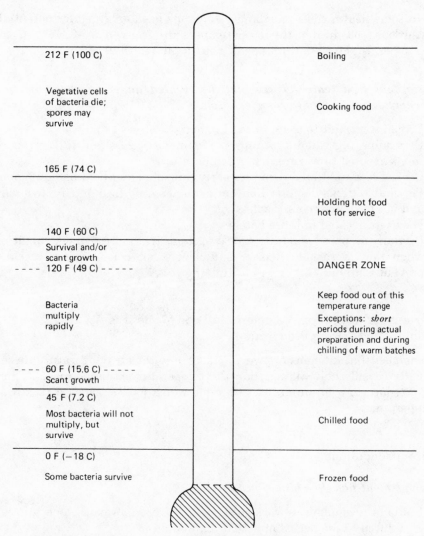

Figure 25. Food temperatures at which bacteria multiply, survive, and die.

range of temperature. As mentioned earlier, the range of 45 to 140 F (7.2 to 60 C) is called the DANGER ZONE because the hazard of bacterial growth is great within that range. The longest period for which food may safely remain in this zone is 4 hours; but food should not be in the 60 to 120 F (15.6 to 49 C) range longer than 2 hours. These time periods are cumulative. This means that every minute that the food has remained at these dangerous temperatures counts toward these 4 hours and 2 hours, respectively.

In some health codes the danger zone has been enlarged for potentially hazardous foods in that the lower temperature is given as 40 F (4.4 C) and the upper as 150 F (66 C). Consult your local health department.

Hazardous food temperatures may be created under a great variety of circumstances. Here are some examples:

- Frozen items left to thaw in the kitchen.
- Perishable ingredients, semiprepared menu items, and fully prepared items exposed to warm kitchen temperatures.
- Cooked food held lukewarm in poorly controlled hot-holding equipment.
- Perishable items displayed on buffet tables, cold food getting too warm and hot foods too cool for safety.
- Food refrigerated in large batches.
- Partially prepared and fully prepared items transported under conditions when they were held neither sufficiently hot nor sufficiently cold to prevent contaminants from multiplying.

Prevent Multiplication of Bacteria in Food While it is Actually Manipulated During Preparation

Remember that kitchens are warm places, and the air temperature there is usually considerably higher than the temperature of other rooms. Therefore, reduce to a minimum the time for which food is exposed to kitchen temperatures.

Recommendations

Management has the obligation to:

1. Plan the employee's work so exposure of food to warm temperatures is reduced to the period of actual manipulation.
2. If at all possible, provide cool rooms for the preliminary preparation of hazardous items involving meat, poultry, fish, and shellfish.
3. Instruct workers to prepare salads and sandwiches from chilled ingredients.
4. Provide refrigeration in the area of food preparation for the temporary storage of partly prepared, hazardous items and for hazardous ingredients such as frozen egg, reconstituted egg, and reconstituted milk.
5. Provide supervision.

The food handler has the obligation to:

1. Follow preparation procedures laid down by management.
2. Endeavor to keep hazardous food items either hot or cold.
3. Cooperate in every way with management in keeping hazardous foods out of the DANGER ZONE—except when food is actually manipulated.

Prevent Multiplication of Bacteria in Food Being Held Hot until Served

Remember that regardless of whether hot food is held in the original cooking equipment, or in special hot-holding equipment, *food temperatures must be maintained at 140 F (60 C) or higher.*

Heated holding equipment is designed for hot-holding, not for heating; therefore, food must be hot, preferably 165 F (74 C) or above, when transferred to the hot-holding equipment.

Recommendations

Management has the obligation to:

1. Provide hot-holding equipment that keeps food at a minimum temperature of 140 F (60 C) and maintain equipment in good working order.
2. Plan food production in such a way that hot food remains in the heated holding equipment for the shortest time feasible.
3. Checks food temperatures frequently with a thermometer in center of batch to ensure that:
   (a) Food is at least 165 F (74 C) when placed in heated holding equipment.
   (b) Food is kept hot (at least 140 F; 60 C) throughout holding.
4. See to it that food that receives much handling followed by light heating in routine cooking is consumed within a short time. This involves *staggered preparation* of items such as:
   (a) Creamed meat, poultry, fish, and shellfish.
   (b) Ground meat patties cooked to the rare or medium stage.
   (c) Crab and fish cakes.
   (d) Croquettes and other lightly-heated products.

The food handler has the obligation to:

1. Fully cooperate with management in the important task of holding hot food hot.

2. Take care to have food hot (at least 165 F; 74 C) all the way through before it is placed in the heated holding equipment.
3. Use good judgment when preparing items by "staggered" operation; keep production just ahead of demand.

Prevent Multiplication of Bacteria in Chilling Food

Remember, that the food temperature that must be reached in order to chill food safely is 45 F (7.2 C) or lower, and that this temperature should be reached as soon as possible, since the food should not be in the DANGER ZONE (140 F to 45 F; 60 to 7.2 C) longer than a total of 4 hours, at the most.

Speed of cooling is affected by many factors:

- Liquid foods cool faster than solid foods.
- Small batches cool faster than large ones.
- Shallow layers cool faster than deep ones (Figure 26).
- Items exposed to low refrigerator temperatures (well below 40 F; 4.4 C). cool faster than those exposed to refrigerator temperatures above 40 F (4.4 C).
- Sufficient refrigerator space should be provided to accommodate the peak load of warm food to be chilled quickly. Refrigerator temperatures should remain at the desired low level throughout.

Cooling food at room temperature is a dangerous practice because foods will cool slowly and temperatures inviting bacterial growth will prevail for a long time. Changes for contamination add to the danger.

The notion is false that food placed in a refrigerator while warm will turn sour because "heat is sealed in."

Recommendations for Chilling

Management has the obligation to:

1. Provide refrigeration equipment sufficient for fast chilling to safe temperatures of all foods requiring refrigeration.
2. See to it that specific refrigerators are used only for the food designated for those refrigerators; for example, meat separated from other items and cooked items separated from raw foods.
3. Provide refrigeration in locations where it can be easily reached by employees; only then *will* it be used.
4. See to it that refrigerators are in good working order, and that repairs, when necessary, are made promptly. See to it refrigerators

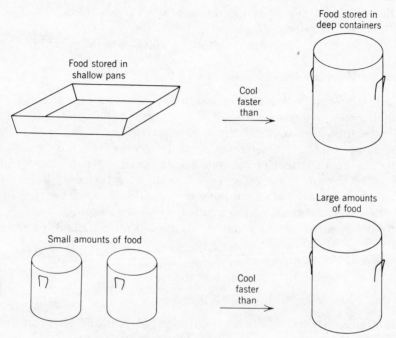

Figure 26. Chilling food.

are kept clean, orderly, uncluttered. Have equipment checked at routine intervals by a reliable serviceman.

5. Provide ample shelving and portable wire racks to avoid stacking of containers. Stacking causes contamination and interferes with the circulation of air.

6. Provide floor racks for walk-in coolers; food should not be stored on floor.

7. See to it that each refrigerator can handle its designated load and that the refrigerator temperature remains low and safe at peak load. Several thermometers should be installed in each refrigerator and read daily; if only one thermometer is used, it should be so installed that it registers in the warmest area of the refrigerator.

8. Provide additional refrigeration if persistently high temperature readings indicate that refrigeration facilities are insufficient.

9. Provide equipment for rapid precooling food outside the refrigerator.

10. Set up appropriate and efficient precooling procedures. The combined cooling time to 45 F (7.2 C) outside and inside the refrigerator should be achieved within 4 hours at the most, and preferably in less time.

11. Set up guidelines for refrigeration of the various items.
12. Provide supervision.
13. Give directions for length of storage, and use of leftovers.

The food handler has the obligation to:

1. Fully cooperate with management in the important task of cooling food rapidly to safe temperatures.
2. Protect food from contamination during cooling.
3. Avoid opening the refrigerator doors unnecessarily often.
4. Avoid allowing the door to remain open for extended periods of time.
5. Keep refrigerator in good order and clean up spills immediately.
6. Refrain from stacking containers on top of each other because of the danger of contamination from soiled container bottoms.
7. Precool food, whenever required, by methods prescribed by management.
8. Use leftovers as directed.

Foods that Need to be Chilled

All potentially hazardous items require rapid chilling and continuous cold-holding whether they are raw, partially or fully prepared in advance, or leftovers.

Hot leftovers that tolerate reheating to boiling should be brought to a boil to destroy contaminants, then chilled rapidly and refrigerated. This rule applies in particular to items that come back from the counter, buffet tables, patient floors, and the like.

GUIDELINES FOR CHILLING FLUID ITEMS AND FLUID ITEMS THAT WILL THICKEN ON COOLING

Small Batches

Distribute food into 1 or 2 gallon (3.8 or 7.6 liter) containers and refrigerate. Do not cool food at room temperatures, except possibly for the first half hour after the heat is turned off. Use shallow pans for puddings and other items that thicken on cooling. Allow to cool uncovered for the first 30 minutes, then cover.

Large Batches

Because heat leaves large batches slowly, precool food to approximately 80 F (27 C); use agitation to hasten cooling. Refrigerate promptly.

- Use a cooling kettle with slowly rotating scraper-agitation. Or use stockpots (2 or 4 gallons; 7.6 or 15.2 liters) immersed in cold running water; agitate with a suitable utensil every 20 to 30 minutes to mix contents well. Refrigerate promptly.
- Refrigerate food in 2 or 4 gallon (7.6 or 15.2 liter) batches. Use shallow pans for puddings and other items that thicken on cooling; or (if desired) distribute food into individual serving dishes, set on trays, set trays on racks, and refrigerate.
- Creamed foods and other items that are mixtures of fluid and solid foods must be handled gently when agitated, lest the solid pieces break up and the consistency of the item be marred.

GUIDELINES FOR CHILLING SOLID ITEMS

- Plan for large roasts, baked hams, and turkeys to be done just before being served; they cool slowly, reheat slowly, and do not lend themselves to advance preparation because of danger from bacterial growth.
- Leftover large roasts, hams, and turkeys will cool faster if cut into smaller portions, sliced (roasts, hams), or disjointed (turkeys), the slices and pieces being then arranged in shallow layers on trays. This practice has its limitations, due to undesirable changes in quality during reheating.
- Always cool meat separately from gravy. For cooling of gravy, see the guidelines for chilling fluid items.
- Whole steamed or simmered turkey cools slowly. Therefore, cut turkeys in half lengthwise prior to cooking. Cool in single layers. Better still, disjoint them and cool in small pieces, single-layered.
- Debone meat and poultry as soon as it is sufficiently cool for handling; bones interfere with cooling. Meat and poultry may be deboned before cooking. This has the advantage of (a) permitting faster cooling and (b) reducing the danger of contamination, which comes from handling the cooked meat.
- Suggestions for safe refrigeration of various hazardous menu items and precautions in the use of leftovers are given in Part Three, Section G.

Length of Chilled Storage

Facts to remember:

1. Storage does not improve cooked food; the contrary is true.
2. Even when food is held at properly low temperature 40 F (4.4 C) or below cold-loving bacteria may spoil the food. Dangerous bacteria may multiply, although slowly.

3. Chilled menu items prepared in connection with ready-prepared (cook/chill) and assembly-serve foodservice systems are stored in the chilled condition for various lengths of time that must be carefully worked out to assure the microbiological and culinary quality of the items. In these systems, the cooked food is chilled, then refrigerator-stored, with the duration of storage varying from one to several days. (For a discussion of these systems see Longrée, Chapter XIV.*)

4. Leftovers are usually subjected to various cycles of holding (heating—hot-holding—chilling) and are likely to be subjected for many hours to temperatures at which bacteria may multiply. In leftovers that are reheated before they are stored, the majority of bacteria are killed. Still, all leftovers are poor sanitary risks, especially cooked items made with meat, poultry, eggs, milk, and fish or other seafood. Cream-filled pastries are also highly perishable and should be served on the day they were prepared.

Recommendations

- Plan to have no or little food left over. Leftovers can be wasteful and are a bacteriological hazard.
- In a well-managed foodservice operation, highly perishable cooked foods are used within the day of their preparation. Check for leftovers after each meal and plan their immediate use.
- Protect food from contamination during storage at all times.
- Daily check refrigerators for sanitary condition, spills, orderliness, and correct temperature.

For handling of specific items, see Part Three, Section G.

Chilling of Displayed Food

- Cold foods to be consumed must be kept chilled while displayed. This includes fruit juices and other cold foods that are not hazardous from a public health point of view.
- Examples of menu items that are potentially hazardous and must be constantly kept chilled for the purpose of bacteriological safety are milk, milk drinks, cream, yoghurt and other perishable dairy products; seafood cocktails; canapes; hors d'oeuvres; cold cuts, meat, poultry, sausage, fish, eggs; salads and sandwiches made with proteinaceous ingredients; potato salad; cold platters; custards; puddings; cream fillings and cream-filled pastries, pies, and cakes; meringued pies; and desserts made with milk

*See the Readings for Part Three for complete reference.

and/or egg, and many others made with potentially hazardous ingredients.

Prevent Multiplication of Bacteria in Frozen Foods

Remember that freezing does not kill bacteria and that bacteria injurious to health may survive freezer storage for months. Also, the survivors will resume multiplication when food has thawed, namely, when ice crystals have melted.

Recommendations for Handling Frozen Food

1. Purchase frozen food from a reputable dealer; check deliveries and refuse items of improper temperature and those showing signs of defrosting. Do not overstock on frozen food.
2. Move delivered items immediately into a freezer maintained at 0 F (−18 C) or colder. Although a storage temperature of 0 F (−18 C) is satisfactory, the freezer temperature is apt to rise when the door is opened frequently, and when large amounts of warm food are introduced.
3. In frozen items that can be cooked in the frozen state, the many dangers connected with defrosting are eliminated. When frozen items must be defrosted, they should not be allowed to warm up beyond 45 F (7.2 C).
4. Frozen menu items prepared in connection with ready-prepared (cook/freeze) and assembly-serve food systems must be subjected to strict time-temperature control during storage, heating and final assembly and/or heating. (See Longrée, Chapter XIV*).
5. *Defrosting Procedures*
   Fresh Frozen Foods Served Unheated
   Fruit Juice Concentrates: Reconstitute from frozen state and chill at once.
   Fruit: Defrost in original container in the refrigerator, or (in a tightly sealed, moisture-proof container) in cold running water.
   Fresh Frozen Foods to Be Heated
   Fruits: Defrost just enough for convenience in handling; defrost in a refrigerator or in a tightly sealed moisture-proof container in cold running water. Fruit, defrosted at room temperature, tends to turn mushy and discolors.

*See the Readings for Part Three for complete reference.

Vegetables: Do not defrost, except corn on the cob, spinach, asparagus, and broccoli.

Potentially hazardous foods that need to be thawed:*

(a) In a refrigerator not warmer than 45 F (7.2 C).

(b) Under safe (potable) *running*, cool water; no warmer than 70 F (21.1 C).

(c) In a microwave oven only when the food will be immediately transferred to conventional cooking facilities as part of a continuous cooking process, or when the entire, uninterrupted cooking process takes place in the microwave oven.

(d) As part of the conventional cooking process.

Examples–Fish, Shellfish, Meats (Small Cuts), and Poultry (Small Cuts): Do not defrost unless breaded or covered with batter. If you wish to defrost, place items in the refrigerator, or enclosed in tightly sealed, moistureproof bags in *cold* running water.

Large Roasts, Whole Turkeys, or Frozen Egg: Always defrost in the refrigerator. Meat and turkey, if enclosed in tightly sealed, moisture-proof bags, and eggs, may also be defrosted in *cold* running water.

Precooked Frozen Items: Defrosting is part of the reheating process that can be done in conventional, convection or microwave ovens. (See also Longrée, Chapters XII, XIII, XIV).

Management has the obligation to:

1. Set up detailed guidelines (Worksheet) concerning defrosting, which should state:

   What food is to be defrosted.

   When to remove it from the freezer.

   Where it should be defrosted.

   How it should be defrosted.

2. Provide the necessary refrigerator space and equipment.

3. Provide supervision.

SAMPLE WORKSHEET
Directions for Defrosting Frozen Foods

Description of food	To be defrosted: Yes No	When to be removed from freezer	Place of defrosting	Additional directions and special precautions

*U.S. Dept. of Health, Education and Welfare, Public Health Service, Food and Drug Administration. *Food Service Sanitation Manual.* See Readings, Part Three.

The food handler has the obligation to:

1. Follow instructions.
2. Report to the supervisor,, if the instructions do not work well, that defrosting times may be "off" and cooperate with the supervisor in improving directions.
3. Carefully check items that are allowed to defrost at room temperatures; should these thaw ahead of time when they are to be cooked, transfer them to a refrigerator for temporary safe holding.

SECTION G

Techniques for Handling Some Specific Hazardous Menu Items

In Tables 8 through 22, recommendations are made for the safe handling, heating, hot-holding, cooling, and cold storage of certain menu items. Precautions in the use of leftovers are also listed.

The menu items selected are those that are frequently mentioned in connection with food-poisoning outbreaks; they are the so-called potentially hazardous items. Their preparation requires the strictest sanitary care at every step.

Table 8. Safety of Creamed Items

Safe Preparation	Safe Heating and Hot-Holding	Safe Refrigeration	Precautions in Use of Leftovers
Use safe, wholesome ingredients; do not use leftovers of doubtful age and wholesomeness.	In general, it is better management and safer to hold sauce and solid ingredients separately until just before serving time; creamed mixtures are a greater risk when held combined than are the sauce and the solids when held separately; also, danger of curdling and of other undesirable changes is lessened.	Creamed foods are hazardous; refrigerate them promptly and quickly.	Heat to at least 165 F (74 C) before serving; this heat treatment will kill many contaminants except heat-resistant spores, and toxins of staphylococci.
Work with clean hands, cutting boards, knives, and other equipment.		Precool thin mixtures (in original cooking utensil), using cold running water; thick mixtures in shallow pans set in ice; if amounts of leftovers are small (1 gallon; 3.8 liters), refrigerate without precooling.	It is almost impossible to predict when the food becomes unsafe; its entire history influences keeping quality and safeness.
Work speedily and efficiently; do not expose foods to contaminants and warmth any longer than absolutely necessary.		Store in the refrigerator in a covered container; use soon, within 24 hours.	Noticeable signs of spoilage (slime, mold, or souring) indicate the presence of microorganisms that may or may not be harmful; it may indicate mishandling and mismanagement.
If ingredients that are to be added to sauce are prepared in advance, refrigerate them until the time for mixing them with the cream sauce has come.	Creamed foods do not conduct heat well; therefore, have sauce as hot as feasible when adding the solids (meat, egg, seafood, etc.).		Absence of spoilage signs does not guarantee safeness.
	Reheat combined mixture to 165 F (74 C) or higher.		DISCARD SUSPECT ITEMS
	Hold mixture above 140 F (60 C) until ready to serve.		

Table 9. Safety of Egg-Thickened Desserts: Stirred Custards, Cream Fillings, Puddings, and Cream-Filled Baked Goods

Safe Preparation	Safe Heating	Safe Refrigeration	Precautions in Use of Leftovers
Use sound-shelled fresh eggs or pasteurized (*Salmonella*-free) processed egg.	Prepare a suspension of one fourth of the cold milk, starch and egg, add to hot milk, and allow to thicken. Mixture is done when raw starch flavor has disappeared.	*Stirred (soft) custards:* pre-cool to 80 F (26.7 C) with the aid of cold running water, under agitation. Refrigerate, covered, in 2 to 3 gallon (7.6 to 11.4 liter) batches.	It is almost impossible to predict accurately when the food becomes unsafe; its entire history influences keeping quality and safeness.
When using shell eggs, prevent pieces of shell from falling into the broken-out egg.	Whenever feasible, cooked mixtures should be placed in cakes, shells, crusts or other baked goods while still hot (at least 140 F; 60 C). Fillings not to be used immediately following preparation should be refrigerated at once and kept cold.	*Fillings and puddings:* pre-cool like custard or refrigerate at once, covered, in shallow pans.	Noticeable spoilage signs (slime, mold, or souring) indicate the presence of microorganisms that may or may not be harmful. They indicate food mishandling and mismanagement.
Break each egg into a small dish and inspect for signs of spoilage, as one spoiled egg may ruin an entire batch of eggs; discard suspect eggs.		Mixtures should be at 45 F (7.2 C) or colder within 1 hour.	Absence of spoilage signs does not guarantee safeness.
Prevent contamination of broken-out shell egg, frozen egg, or dried egg; sources of contamination are hands, soiled ladles, and other equipment and utensils.		Cream-filled and custard-filled cakes and pastries must be kept under constant refrigeration.	DISCARD SUSPECT ITEMS
Prevent contamination of the cooked product.		Use stirred custards, cream fillings, puddings, and cream-filled baked goods, within 24 hours.	

Table 10. Safety of Meringued Pies

Safe Preparation	Safe Heating	Safe Refrigeration	Precautions in Use of Leftovers
Use sound-shelled fresh eggs or pasteurized (*Salmonella*-free) processed eggs.	Bake in moderate oven 325 to 350 F (163 to 177 C) for 16 or 20 minutes. Reason: meringues baked quickly at higher temperatures become brown on top before the portion near the filling is done and are not thoroughly heated.	Keep finished pies under constant refrigeration. Use within 24 hours.	It is almost impossible to predict accurately when the food becomes unsafe; its entire history influences keeping quality and safeness.
Prepare crust.			Noticeable spoilage signs (slime, mold, or souring) indicate the presence of microorganisms that may or may not be harmful. They indicate mishandling and mismanagement.
Prepare filling.			
Prepare meringue.			
Pour filling into crusts while filling is still hot, at least 158 F (70 C), and add meringue immediately.			Absence of spoilage signs does not guarantee safeness.
Bake pies while filling is still hot.			DISCARD SUSPECT ITEMS

Table 11. Safety of Hollandaise

Recommendations
Use high-quality, sound-shelled fresh eggs or pasteurized processed eggs.
Stagger production. Prepare a new batch every two hours. This sauce cannot be held at safe temperatures without disintegrating in consistency, therefore prepare only small amounts at a time.
Discard sauce that has been held more than two hours at temperatures below 140 F (60 C)

Table 12. Safety of Menu Items Containing Uncooked Egg

Items	Recommendations
Acid items	Use high-quality sound-shelled eggs, or pasteurized (*Salmonella*-free) processed egg.
Mayonnaise and uncooked salad dressing	Prevent contamination of egg mass from pieces of shell, spoiled egg, soiled ladles, soiled hands, and other sources of salmonellae.
	Use clean equipment for preparation and storage of dressing.
	Store, covered, in the refrigerator.
	Do not contaminate the dressing when removing portions from the batch. Use clean utensils.
Nonacid items	
(a) Desserts made with egg foam, such as chiffon-type items and whips	Use high-quality sound-shelled eggs, or pasteurized (*Salmonella*-free) processed egg white.
	Prevent contamination of egg from pieces of shell, spoiled egg, soiled ladles, soiled hands, and other sources.
	Use clean bowls and beaters in whipping, folding, and dipping procedures.
	Refrigerate finished items at once and keep under continuous refrigeration until served.
	Serve within 24 hours.

Table 12.   (Continued)

Items	Recommendations
(b) Egg-enriched nourishment, such as eggnog.	The use of uncooked shell eggs in nourishment for patients is inadvisable because of the danger from *Salmonella*.
	Pasteurized processed egg, although rendered free from *Salmonella* in pasteurization, may well become recontaminated.
	Remember that the very young, the aged, the frail, and the sick are particularly sensitive to *Salmonella*.
	Commercial eggnog mixes made with pasteurized egg may be used with precaution: the powder must be kept in a closed container and protected from contaminants; and clean spoons must be used to remove the portion to be reconstituted into fluid eggnog. Once reconstituted, eggnog is excellent food for bacteria and therefore must be kept chilled.
	Baked custard may be served in place of eggnog if solid nourishment is permitted.
(c) Breading mixture	The batter becomes easily contamined by microorganisms clinging to the items breaded such as cutlets, steaks, chicken parts, seafood.
	At kitchen temperatures the contaminants will multiply readily in batter. Bread crumbs may become contaminated by drippings and/or by food handlers' hands.
	Renew batter frequently, at least hourly.

Table 13. Safety of Soups and Soup Base

Safe Preparation	Safe Heating and Hot-Holding	Safe Refrigeration	Precautions in Use of Leftovers
Use safe, wholesome ingredients; beware of using leftovers of doubtful history. Work with clean hands, use clean cutting boards, knives, and other equipment and utensils.	In general, soups made with stock are boiled or brought to a boil; this is effective in destroying contaminants, although heat-resistant spores of *C. perfringens* may survive. Cream soups, especially those containing seafood and leftovers, should be heated to as high a temperature as possible, at least 165 F (74 C). Do not prepare large amounts of stock and soup in advance unless you can keep them safe, either hot, 140 F (60 C) and above; or cold, 40 F (4.4 C) and below.	Precool in cooling kettle, using agitation; or precool in 2 to 4 gallon (7.6 to 15.2 liters) batches, setting stock pots in cold running water and stirring occasionally (every 15 to 30 minutes). When the steaming stops, refrigerate. Avoid contamination from soiled containers, soiled spoons, hands, dust, etc. Store in refrigerator in covered containers. Use leftover soups as soon as possible, preferably within 24 hours.	With prolonged storage, the number of bacteria will increase; toxin of staphylococci may form. Reheat leftover stock and soup to a boil before serving. It is almost impossible to predict accurately when the food becomes unsafe. Its entire history influences keeping quality and safeness. Absence of spoilage signs does not guarantee safeness; activities of food-poisoning bacteria do not noticeably change appearance, odor, or flavor of contaminated items. DISCARD SUSPECT ITEMS.

Table 14. Safety of Fish and Shellfish Items: General Directions

Safe Preparation	Safe Heating and Hot-Holding	Safe Refrigeration	Precautions in Use of Leftovers
Use safe, wholesome fish and shellfish harvested from unpolluted waters.	Use certified, safe raw materials only. Some shellfish, like oysters and clams, are eaten uncooked or very lightly heated.	Fish and shellfish are to be kept under constant refrigeration except during actual preparation.	It is almost impossible to predict accurately when the food becomes unsafe; its entire history influences quality and safeness.
If frozen and defrosting is necessary, follow approved methods of defrosting.	Serve cooked seafood immediately when desired stage of doneness is reached. Holding these items at safe temperatures for long may toughen them, while holding them at temperatures below 140 F (60 C) invites bacterial growth. Therefore, STAGGER PRODUCTION.	Precool leftover cooked seafood by recommended procedures, then refrigerate.	Noticeable spoilage signs (slime, mold, souring, etc.) indicate the presence of microorganisms that may or may not be harmful, but that indicate mishandling and mismanagement.
Use clean hands, cutting boards, and utensils; use care in deboning and deveining; wash hands frequently.		Use leftovers promptly (within 24 hours).	Absence of signs of spoilage does not guarantee safeness; activities of food poisoning bacteria do not noticeably change appearance, odor, or flavor.
If items are breaded, use special precautions that prevent the build-up of bacteria in the egg mixture. Prepare fresh mixture each hour.			DISCARD SUSPECT ITEMS.
Manage preparation so that seafood does not remain at lukewarm temperatures longer than 1 hour at the most. Keep items under refrigeration except when actually manipulated.			

Table 15. Safety of Gravies

Safe Preparation	Safe Heating and Hot-Holding	Safe Refrigeration	Precautions in Use of Leftovers
Prevent contamination of ingredients as well as finished gravy. Use clean equipment and utensils in preparation.	Boil or simmer gravies and reheat them each time after a chance for contamination has occurred; for example:	Refrigerate leftover gravy at once. Precool in cooling kettle or in 1-gallon (3.8 liter) stock pots set in water or ice; stir occasionally.	Do not add freshly prepared gravy to leftover gravy. Before serving leftover gravy, reheat thoroughly; boiling or simmering is recommended.
Prevent multiplication of contaminants by keeping the gravy hot or cold.	(a) reheat poultry gravy to simmer, or boil, after adding chopped ingredients such as giblets.	Store the gravy in a covered container; use within 1 to 2 days.	It is almost impossible to predict when the food becomes unsafe; its entire history influences keeping quality and safeness.
	(b) reheat meat gravy to simmer or boil after adding juice from cooked roast.		Noticeable spoilage signs (slime, mold, souring, etc.) indicate the presence of microorganisms and indicate mishandling and mismanagement.
	Keep gravy hot (simmering) until serving time. If prepared in advance, cool quickly and refrigerate.		Absence of spoilage signs does not guarantee safeness.
			DISCARD SUSPECT ITEMS.

Table 16. Safety of Beef Roasts

Safe Preparation	Safe Heating and Hot-Holding	Safe Refrigeration	Precautions in Use of Leftovers
Use only meat inspected by the government for wholesomeness.	Cook roasts to an internal temperature of at least 130 F (54 C)*	Before handling meat, wash hands. Use clean sanitized work surfaces, equipment, and utensils. Avoid using hands any more than absolutely necessary.	Avoid contamination and warming of meat whenever it is temporarily removed from refrigeration during storage.
Use separate working surfaces and equipment for raw and cooked meat. If this is not feasible, thoroughly clean and sanitize work tables, chopping blocks, slicers, etc. after they are used for raw meat. (See Part Three, Section B.)	Plan roasting time of rolled roast so that the roast is finished just before serving. This is recommended because of the danger from inside contamination that may occur during deboning and rolling. A contaminant of special importance is *C. perfringens*; its spores survive roasting.	Chill cooked meat separately from juice and gravy.	It is almost impossible to predict when the food becomes unsafe; its entire history influences keeping quality and safeness.
Thoroughly wash hands after handling raw meat.		Bone-in, cooked roasts may be refrigerated in bulk. Refrigerate promptly.	Noticeable signs of spoilage (slime, mold, souring, etc.) indicate mishandling and mismanagement.
Keep meat cold until ready to be roasted.	Ideally, the period of hot-holding should not exceed two hours.	Boned cooked roasts cool quickly and more safely if sliced. Arrange slices in shallow pans, cover, and chill at once.	Absence of spoilage signs does not guarantee safeness.
Defrost frozen roasts in a refrigerator.	Rolled roasts to be served cold should be sliced soon after cooking; place slices in shallow pans, cover, and refrigerate at once.	Refrigerate only the amount that can be used in 1 to 3 days; freeze the rest.	DISCARD SUSPECT ITEMS.

*Temperatures recommended in the Food Service Sanitation Manual of the U.S. Dept. of Health, Education and Welfare, Public Health Service, Food and Drug Administration. See Readings for Part Three.

Table 17.   Safety of Cooked Pork and Ham

Special Hazards	Recommendations
Danger from trichinae.	Roast fresh pork to at least 150 F (66 C), preferably to 170 F (77 C), the gray stage, to be safe.
	Bake hams to at least 150 F (66 C), including tenderized items.

Table 18. Safety of Poultry Items*

Safe Preparation	Safe Heating and Hot-Holding	Safe Refrigeration	Precautions in Use of Leftovers
Use only poultry government inspected for wholesomeness.	Poultry is not acceptable unless cooked to doneness. Fortunately, cooking to the well-done stage eliminates the danger of salmonellae surviving the cooking. However, stuffed large turkeys are not safe.	Cool and store meat separately from gravy.	It is almost impossible to predict when the food becomes unsafe; its entire history influences keeping quality and safeness.
Defrost frozen poultry in a refrigerator.		Refrigerate leftovers promptly.	
Keep poultry cold until ready to be prepared.		Leftover whole turkey should be deboned and meat placed in clean shallow pans or on trays. Be sure working surfaces are clean and sanitized. Cover, and refrigerate no longer than 1 to 2 days.	Noticeable signs of spoilage (slime, mold, souring, etc.) indicate mishandling and mismanagement.
Thoroughly wash inside of poultry carcass before roasting. Wash giblets.	If giblets are used in gravy, chop cooked giblets on sanitized board, add to gravy, and boil gravy.		Absence of spoilage signs does not guarantee safeness.
Use separate working surfaces and equipment for raw and cooked poultry. If this is not feasible, thoroughly clean and sanitize work surfaces and equipment that were previously used for raw poultry (see Part Three, section B); only then use them for cooked items.	Total warm holding of poultry and gravy should not exceed 2 hours.		DISCARD SUSPECT ITEMS.
Thoroughly wash hands after handling raw poultry.			

*See also Table 19, "Safety of Deboned Poultry."

Table 19. Safety of Deboned Poultry*

Safe Preparation	Safe Heating and Hot-Holding	Safe Refrigeration	Precautions in Use of Leftovers
Poultry may be deboned while raw.	Meat is safe when cooked to doneness; the dangers arise from handling the cooked meat and holding it at temperatures at which bacterial growth is possible.	To refrigerate poultry, arrange meat in shallow layers, whether it is the entire bird, large pieces, or chopped meat.	Use cooked poultry the same day, if at all possible.
Cook poultry to doneness.			Menu items made with much-handled poultry meat, such as salads and creamed items (à la king) are frequently the cause of food-poisoning outbreaks.
Move cooked poultry to shallow pans or trays to cool in a cold place away from dust and other chances for contamination.		Cover and refrigerate no longer than 1 to 2 days.	
Unless deboned previously, remove meat from bones as soon as meat is sufficiently cool to handle.			It is almost impossible to predict when the food becomes unsafe; its entire history influences keeping quality and safeness.

Do not allow the process to drag along; the job should be completed in 20 to 30 minutes.

Place meat in shallow pans or on trays to cool fast and chop when cool; or chop at once, place meat in shallow pans or on trays and cool. Work with meticulously clean hands. Wash hands before starting work and each time you touch anything other than the meat.

Use clean and sanitized chopping board. If possible, use board not previously used for raw meat.

Noticeable signs of spoilage (slime, mold, souring, etc.) indicate mishandling and mismanagement.

Absence of spoilage signs do not guaranee safeness.

DISCARD SUSPECT ITEMS.

*Prepared for use in creamed dishes, salads, sandwiches, and other combination dishes. See also Table 8, "Safety of Creamed Items"; Table 20, "Safety of Salads"; Table 21, "Safety of Sandwiches"; and Table 22, "Safety of Hot Menu Items."

Table 20. Safety of Salads (Potato, Meat, Poultry, Egg, and Seafood)

Safe Preparation	Safe Heating and Hot-Holding	Safe Refrigeration	Precautions in Use of Leftovers
Use safe, wholesome ingredients; especially beware of leftovers of doubtful history or old leftovers.	No heat can be applied to the prepared ingredients or the finished salads.	Keep hazardous prepared ingredients and salads chilled at all times except during actual preparation.	Potato, meat, poultry, egg and seafood salads have been indicted many times in outbreaks of foodborne illness.
Have the ingredients cold.		Heat transfer is slow in salads; thus, cooling in large batches represents a special hazard. Do not pile salad high in a large container; cool in shallow pans to speed heat transfer; cover salads.	Leftover salads exposed to warm temperature several hours before refrigeration are hazardous, as are salads repeatedly removed from refrigerator and exposed to warm temperatures.
Work with clean hands and equipment when preparing ingredients and combining them.		Use as soon as possible, certainly within 24 hours.	It is almost impossible to predict when the food becomes suspect; its entire history influences keeping quality and safeness.
Add acid ingredients to main ingredients at early stage of preparation.			Noticeable signs of spoilage (slime, mold, souring, etc.) indicate mishandling and mismanagement.
Use about ⅓ of the salad dressing (use it as a marinade) immediately after the potatoes, meat, poultry or seafood have been peeled, sliced, chopped, deboned, etc.; the more acid dressing is used, the greater its inhibitory effect on the bacteria. Refrigerate marinated mixture and add remaining ingredients just before serving.			Absence of spoilage signs does not guarantee safeness.
			DISCARD SUSPECT ITEMS.

Table 21. Safety of Sandwiches Made With Meat, Poultry, Egg, and Fish

Safe Preparation	Safe Heating and Hot-Holding	Safe Refrigeration	Precautions in Use of Leftovers
Use safe, wholesome ingredients; beware of leftovers of doubtful history.	Hot sandwiches should be thoroughly heated, or made with thoroughly heated ingredients (meat, gravy, etc.) and served at once.	Do not stack high; spread them in shallow layers on trays to a height of 3 to 4″ (7.5 to 10 cm).	Sandwiches that contain hazardous ingredients and are exposed to warm temperatures for hours are suspect. DISCARD.
Prepare sandwiches from chilled ingredients, except butter.	Heat cannot be applied to cold sandwiches.	Keep under strict refrigeration at all times; this includes display area on cafeteria counters.	
Work with clean hands and equipment.		Use within 24 hours.	
Stagger production; place finished sandwiches on shallow trays no more than approximately 3 to 4″ (7.5 to 10 cm) high; refrigerate. Start a new batch frequently.			
If feasible, use acid to discourage bacterial multiplication; place pickle, salad dressing, mayonnaise, prepared mustard, horseradish, etc. next to meat or other hazardous fillings.			
Wrap sandwiches for display.			
Chill finished sandwiches rapidly; keep chilled until served.			
On catered, wrapped sandwiches, show the date of preparation.			

Table 22. Safety of Hot Menu Items Made with Ground, Diced, Chopped, Flaked, Cubed, or Otherwise Minced Meat, Poultry, Egg, Fish, and Shellfish.

Safe Preparation	Safe Heating and Hot-Holding	Safe Refrigeration	Precautions in Use of Leftovers
Use fresh, wholesome ingredients. Beware of leftovers.	Group I (meat, poultry and seafood patties, cakes, croquettes and hamburgers cooked to the medium rare or rare stage):	Plan production of these hazardous items in such a way that there are few or no leftovers.	Leftovers of *items made with fresh ingredients:*
If leftovers must be used, take only those that were treated in a sanitary manner, chilled rapidly, and refrigerator-stored for no longer than 24 hours.	Keep items under rerigeration until heated.	Leftovers of menu items in which leftovers have been used are particularly hazardous. DO NOT REUSE.	Use within 24 hours. DISCARD SUSPECT ITEMS. Leftovers of *items made with leftover ingredients:* DISCARD.
Work with clean hands; use clean work surfaces, equipment, and utensils.	Stagger the heating in such a way that each batch is consumed within one half hour of having finished cooking, or sooner. Warm holding for extended periods is hazardous.		
Work rapidly. Unless items are cooked at once, place them in a refrigerator and bring them out as needed.	*Group II (casseroles and meat and poultry pies):*		
Treat with respect all egg mixtures used in breading and battering. Make up fresh mixture every hour or so to prevent build-up of food-poisoning bacteria that may stem from the meat, the worker's hands, soiled equipment, or all of these.	Keep items under refrigeration until heated. For thorough heating, cook in moderate oven until the mixture boils. *Do not* heat under the broiler. *Group III (turkey stuffing):* Bake separately to 165 F (74 C) internal temperature.		

Important Facts in Brief for Part Three

- Sanitation of the physical plant involves:
  1. Cleanability of walls, floors, and ceiling.
  2. Proper sewage disposal and plumbing; presence of floor drains where required because of spillage or method of cleaning floor.
  3. Good ventilation.
  4. Good lighting.
  5. Adequate hand-washing facilities.
  6. Sanitary dressing rooms or locker room and sanitary toilet facilities.
  7. Sanitary garbage disposal.
  8. Control of rodents and insects.
  9. Separate storage area for cleaning materials and utensils.
  10. Regular and adequate cleaning of the facility.

- The primary objective of any cleaning operation is to remove soil and control bacteria. This is necessary to prevent food spoilage, reduce health hazards, and control odors.
- The main cleaning agent is water. Detergents and friction help water do a better job.
- Clean water at 170 F (76.6 C) for one minute is necessary to rinse away soil and sanitize the item being cleaned.
- To plan a successful cleaning operation, the cleaner must consider the nature of the water, the type of soil, the corrosion resistance of the material being cleaned, the type of cleaner, and the condition of the soil.
- Cutting, dicing, and mixing equipment should be easy to dismantle so parts that come in contact with food can be properly cleaned and sanitized.
- The sanitary concern of primary equipment such as ovens, fryers, and broilers is one of odor control, esthetics, and equipment efficiency.
- Dishwashing or pot-and-pan machines require regular cleaning if dishes and utensils are to be properly cleaned and sanitized.

- Transportation and dish-storage equipment needs to be cleaned regularly to avoid recontamination of clean dishes and pans.
- A written cleaning procedure is a useful training tool and serves as a standard for judging the quality of performance.
- A checklist provides a handy reference of items that need to be cleaned and the frequency of cleaning required.
- Safety is an integral part of sanitation.
- "Accidents don't happen; they are caused," therefore they can be prevented, but only through the concerted effort of management and the workers.
- By law: Safety on the job is everybody's responsibility!
- Some of the items that the compliance officer will check during an inspection are:

  1. The accessibility of fire extinguishers, and their readiness for use.
  2. Guards on floor openings, balcony storage areas, and receiving docks.
  3. Proper and adequate handrailing on all stairs.
  4. Guards for machines and machine parts.
  5. Electrical grounding of equipment and tools.
  6. General housekeeping with areas clean and in good repair.
  7. Passageways lighted and clear of obstruction.
  8. Compliance with OSHA posting requirements.
  9. Compliance with OSHA recordkeeping.
  10. Proper use of extension cords.
  11. Exits marked and kept clear.

- For a sanitation program to be effective, one should start with a sound food supply. To assure wholesomeness, it is of greatest importance to specify exactly what one wants.
- Some foods must be considered potentially hazardous because of their composition and/or the handling they receive. The definition of hazardous foods as given by the U.S. Public Health Service reads as follows:

Potentially Hazardous Food shall mean any perishable food which consists in whole or in part of milk or milk products, eggs, meat, poultry, fish, shellfish, or other ingredients, capable of supporting rapid and progressive growth of infectious or toxigenic microorganisms.

- The health of the food handler plays an important part in sanitation. Food handlers can be sources of bacteria causing illness through transmission of disease germs or through food poisoning.
- Management and employees should cooperate in the effort to keep ill

employees from working with food, and with equipment and utensils used in preparing and serving food.

To safeguard the wholesomeness of finished menu items at the time they are served, the following rules need to be observed:

1. Start out with wholesome food material.
2. Remove dirt, soil, and chemical residues by washing thoroughly.
3. Protect food from contamination with poisonous substances, soil, and disease-producing bacteria throughout its preparation, storage, and service.
4. Thoroughly heat those foods that can be expected to harbor illness-producing microorganisms whenever feasible.
5. Prevent multiplication of microorganisms in the food and on equipment used in food preparation, storage, and service.

Study Questions for Part Three

1. What is the food operator's responsibility to the customer?
2. What is the primary objective of any cleaning operation? Why is this objective necessary?
3. What is the main cleaning agent? What functions do detergents and friction serve?
4. Under what conditions is water a satisfactory sanitizer?
5. What features are important in purchasing food-preparation equipment?
6. Why are chopping, grinding, and mixing machines considered to have a high potential for causing microbial contamination of food?
7. What information should a cleaning schedule give?
8. What law mandates safe and healthful working conditions?
9. What government department is responsible for administering this law?
10. What are some items that the OSHA compliance officer will look for in inspecting a foodservice operation?
11. What are the four types of violations considered in an OSHA inspection and which are most important for foodservice operators and workers?
12. What government agencies are responsible for safeguarding the wholesomeness of food?
13. What food products are considered potentially hazardous?
14. What is meant by the DANGER ZONE in food handling?
15. What time-temperature relationships must be observed to pasteurize milk?
16. What are the general rules for purchasing a sound food supply?
17. What are the general rules for checking deliveries and storage of food?
18. What are management's responsibilities in "safe" food handling? What are the employee's responsibilities?

Readings for Part Three

American Society for Hospital Foodservice Administrators. *OSHA Reference for Food Service Administrators.* American Hospital Association, Chicago. 1976.

Bjornson, B. F., H. D. Pratt, and K. S. Littig. *Control of Domestic Rats and Mice.* U.S. Department of Health, Education and Welfare, Public Health Service. PHS Publ. No. 563. Revised 1968.

Center for Disease Control. *Foodborne and Waterborne Disease Outbreaks.* Annual Summary. U.S. Dept. of Health, Education and Welfare, Public Health Service. (Published annually).

Federal Register. *Occupational Safety and Health Standards.* 39:23502-23828. June 27, 1974.

Kotschevar, Lendal H. *Quantity Food Production.* 3rd. ed. Cahners Books, Boston, 1974.

Kotschevar, L. H. *Quantity Food Purchasing.* 2nd ed. Wiley, New York, 1975.

Kotschevar, L. H. and M. Terrell. *Food Service Planning: Layout and Equipment.* 2nd ed. Wiley, New York, 1977.

Longrée, K. *Quantity Food Sanitation.* 3d ed. Wiley, New York. 1980. Chapters VIII through XIV.

National Restaurant Association. *Technical Bulletin about Occupational Safety and Health Act.* Chicago. 1975.

Occupational Safety and Health Administration, U.S. Department of Labor. *All about OSHA.* OSHA 2056, April 1976. Available from OSHA (free).

Occupational Safety and Health Administration, U.S. Department of labor. *OSHA Inspections.* OSHA 2098, June 1975. Available from OSHA (free).

Occupational Safety and Health Administration, U.S. Department of Labor. *Workers Rights under OSHA.* OSHA 2253. October 1975. Available from OSHA (free).

Science and Education Administration, U.S.D.A. *Coackroaches, How to Control Them.* U.S.D.A. Leaflet No. 430. Washington, D.C. 1978. (For sale by the Superintendent of Documents, Government Printing Office, Washington, D.C. 20402. Stock no. 001-000-03835-4.)

Terrell, Margaret E. *Professional Food Preparation.* 3rd ed. Wiley, New York. 1979. Pp. 35–45.

U.S. Department of Health, Education and Welfare, Public Health Service, Food and Drug Administration. *Food Service Sanitation Manual.* 1976 Recommendations. DHEW Publication no. (FDA) 78-2081. 1978. (For sale by the Superintendent of Documents, Government Printing Office, Washington, D.C. 20402.)

U.S. Department of Health, Education and Welfare, Public Health Service, Food and Drug Administration. *The Vending of Food and Beverages, Including a Model Sanitation Ordinance.* 1978 Recommendations. DHEW Publication no. (FDA) 78-2091. 1978. (For sale by the Superintendent of Documents, Government Printing Office, Washington, D.C. 20402.)

West, Bessie B., L. Wood, V. Harger, and G. Shugart. *Food Service in Institutions,* 5th ed. Wiley, New York. 1977. Sections: I-3; II-10; III-11 and 12.

PART FOUR

Education and Training in Sanitation of Foodservice Personnel

People can make or break a good sanitation program. The important role that workers and management play in the safe handling of food has been stressed throughout this book. The money, time, and effort that go into a sanitation program are wasted if the people working in foodservice are not aware of or knowledgeable about appropriate sanitary practices.

Sanitation-conscious workers don't just happen. They are developed by providing good leadership, the proper tools to do the job, and continuous training and follow-up to develop safe food-handling habits. Supervision or follow-up recommended sanitation procedures must be carried out daily if the worker is to believe that sanitation is important.

Industrial training may be defined as the art of helping employees to acquire "desirable habits" necessary for the performance of their tasks. Habits result from practice or repetitive doing. Desirable habits result from the practice of the correct procedures; thus, before new employees can start practicing, they must have a clear mental picture of what they are trying to accomplish and know why this is important. By the same token, the trainer must also have a clear mental picture of what he or she wants done and how the job should be accomplished.

Before developing a successful sanitation training program, management needs to evaluate the current sanitation procedures in use. There is little point in expending resources on any training programs if that training results in ineffective sanitation procedures. Top management must be willing to provide the necessary people, tools, and equipment to implement effective sanitation on a continuing basis. A common and justifiable complaint from workers is that they are frequently required to attend training sessions and then are not able to apply what they have learned to their jobs because of the lack of time, money, materials, or equipment.

The teaching ability and work habits of the potential trainers should also be evaluated. Management should not assume that because a supervisor knows how to do a job, the individual does that job in the most effective manner or can teach others to do the job properly. The frequency with which foodservice establishments are cited as being unsanitary would suggest that it is easier to teach the "wrong way" to do a job than the "right way." Too often this is due to the instructor. Usually the trainer teaches the worker to do the job in the same order in which the instructor does it. This may not be the best sequence for the beginner, thus the trainer may confuse rather than teach the worker. Almost always the trainer thinks of the cleaning job in bigger units than the new worker can understand. Hence the instructor overlooks many of the small but essential details the worker must learn in order to do the job successfully. Unless the trainer is freed of other responsibilities during the training period, he or she may not have time to teach the whole job at one time. When training is done in "bits and pieces," the learner never gets a clear mental picture of the total job or the explanation of why the job must be done a certain way. As a result the worker is tempted to develop shortcuts that lead to unsatisfactory cleaning practices. Even more important, if the teacher does not have time to do constructive follow-up of the training, the student may develop undesirable work habits relative to sanitation. Thus the first step in developing a good training program may well be to train the trainer.

The J.I.T.* technique developed by the government during World War II has been used successfully in training. This technique is divided into two parts. The first part gives the key activities of the trainer that are necessary before training starts. The second part gives the four basic steps required in training the worker.

Before training can start, the trainer should develop a training plan and schedule. The training plan answers the question of *who* is going to be trained, *what,* and *where.* The training schedule tells *when* training will be done. Next the trainer should prepare a written job breakdown of the cleaning activity to be accomplished. This job breakdown or cleaning procedure will list the important steps in the activity and the key points necessary for accomplishing each step. Such a cleaning procedure can serve as a teaching tool during the period of instruction, as a reference point for the student during the practice sessions, as the instructor's check sheet during the follow-up, and as an evaluation tool in checking the results. Finally, the trainer will need to get ready to teach by securing the necessary supplies and equipment for the job. The work place must be arranged as it will appear when the worker is responsible for the cleaning. The trainee should be able to stand next to rather than across from the instructor during the demonstration. It is very important that the trainee stand in approxi-

*Job Industry Training.

mately the same position as he or she will when actually doing the job. If the trainee stands across from the instructor he or she will get a mirror image rather than a true image of the procedure. This may result in confusion when the real position is taken in front of the equipment in the practice session.

The four basic steps in training can be outlined as follows:

1. Prepare the employee for instruction.
   (a) Put the worker at ease.
   (b) Create in the student a desire to learn the specific task by explaining the job.
   (c) Stress:
      1. The importance of good sanitation relative to this job.
      2. Safety.
      3. Operational efficiency.
2. Present the job.
   (a) Set a clear pattern for the learner to follow by using the cleaning procedure.
   (b) Explain and demonstrate one step at a time.
   (c) Stress the key points.
   (d) Discuss and show the minor details that are necessary to accomplish the job.
   (e) Omit all material that is not important for the job.
3. Practice period.
   (a) Have the student do the job.
   (b) Have the student tell and show what is going to be done and why.
   (c) Correct errors and omissions as the student makes them.
   (d) Continue to practice until the student is able to give back to the teacher everything that the teacher gave the learner in step 2.
4. Follow through or test.
   (a) Put the worker on his or her own.
   (b) Encourage questions.
   (c) Come back to the work place frequently to check progress or problems.
   (d) Tell the worker how he or she is doing.

The instructor must have a clear picture of every detail; otherwise, he or she cannot expect to pass that picture on to the student. The instructor must also have skill in handling the equipment, if the learner is to have any confidence in the instructor. The J.I.T. motto is: If the pupil hasn't learned, the teacher hasn't taught. Worker performance is the criterion used to determine the quality of the instruction.

In the past, most of the training programs in sanitation have been directed at foodservice employees. These programs, however, have failed to keep pace with the needs of the industry due primarily to:

- The number of young people that enter the industry each year.
- The high turnover rate among food handlers already in the industry.
- The increase in the number of foodservice operations and the number of people being served.
- Technological changes in food processing, food microbiology, foodservice equipment, and cleaning supplies.

Therefore the need to restudy the problem became self-evident. Participants at the 1971 National Conference on Food Protection sponsored by the American Public Health Association, the Food and Drug Administration, and representatives of the foodservice industry recognized that the foodservice manager was the key to a clean and sanitary operation. Thus, one of the recommendations that came from this conference was that the focus of the sanitation training programs should be redirected toward management. The foodservice manager should be trained in good sanitary techniques and certified after his or her qualifications to protect the public by operating a clean and sanitary foodservice business are demonstrated.

In 1974, the National Institute for the Foodservice Industry (NIFI) in cooperation with National Sanitation Foundation, the National Environmental Health Association, the National Restaurant Association, and the Food and Drug Administration developed a management course: *Applied Foodservice Sanitation.* All students who complete the course and pass an examination administered under monitored conditions receive a NIFI Certificate of Completion.

At present about 35 states have some form of foodservice manager's certification program. In some states these programs are mandatory and in others they are voluntary. Regardless of the voluntary or mandatory status of these programs, two trends are evident: first, a tightening of sanitation regulations is spreading across the country and is supported by an increasing number of health jurisdictions and the industry's consumers; second, the focus of sanitation training is changing from the food handlers to managers. The consensus is that managers must know sanitation and proper food protection as well as they know other parts of their job; that they need this knowledge so they can train the food handlers working for them; and that they can integrate food sanitary techniques into their daily work schedules and can assume responsibility for some self-inspection of their operations every day (See Appendix B.)

Whether we are talking about sanitation training for the foodservice worker or management training and certification, it is important to remember that the education programs must be continuous if they are to meet the needs of the industry. At present, the training and certification of foodservice managers shows great promise for improving food protection in the foodservice industry. Whether that promise is realized will depend on whether management training and certification is treated as the continuous process it should be or is merely offered as a one shot deal.

Study Questions for Part Four

1. What does management have to provide to implement effective sanitation on a continuing basis?
2. What are the two parts of the Job Industry Training (J.I.T.) technique?
3. What does the trainer have to do before training can start?
4. Outline the four basic steps in the J.I.T. technique and explain how each step should be carried out.
5. Why should the trainee stand next to the trainer rather than across from him during a cleaning demonstration?
6. What is the J.I.T. motto?

Readings for Part Four

Clingman, C.D. "Foodservice Manager Certification—The NIFI Program." *J. of Food Protection* 40(3):196–7. March 1977.

Davis, Sidney A. "Sanitation Training for Foodservice Managers—a Must." *J. of Food Protection* 40(3)198–99. March 1977.

Hall, C.G. "National Program for Foodservice Sanitation." *Food Technology* 31(8):72. August 1977.

Longrée, Karla. *Quantity Food Sanitation.*, 3rd ed. Wiley-Interscience, New York. 1980. Chapter XV.

Standard Brands, Inc., "How to Train—Job Instructors Guide, Tested Management Techniques." Standard Brands, Inc., New York.

Stauffer, Lee D. "From Pots and Pans to Large Appliances: Equipment and Food Safety." In *Sanitation in Hospital Food Service,* reprinted from *Hospitals, Journal of the American Hospital Association, August 16, 1964.*

U.S. Department of Health, Education and Welfare, Public Health Service, Food and Drug Administration. *Food Service Sanitation Manual 1976.* DHEW Publication No. (FDA) 78-2081. 1978.

Visual Aids

"Don't be a Missing Link," filmstrip on safety for the foodservice industry prepared by the National Safety Council and the National Restaurant Association. Chicago, Ill. 60611.

Package FS-1 Hospital Food Service Series—Sanitation.

  Five film strips and records that accompany the film strips.

  Trainex Corporation, P.O. Box 116, Garden Grove, California. 92642

"Protecting the Public" tells your employees:

  Part one—the personal side.

  Part two—food protection.

  Part three—establishment and equipment sanitation.

  The Educational Materials Center, National Restaurant Association, 2600, One IBM Plaza, Chicago, Ill. 60611

"Protecting the Public"—Health.

  Slides/cassette featuring the 1976 FDA Model Foodservice Ordinance.

  The Educational Materials Center, National Restaurant Association, Suite 2600, One IBM Plaza, Chicago, Ill. 60611

Sanitation Series

  FS 122-125, 134, 147, 150

Safety Series

  FS 109-113, 130

National Education Media, Inc., 15760 Ventura Boulevard, Encino, Calif, 91436.

  Catalog on request.

APPENDIX A:

Tables I to VII

Table I. Transmission, by Way of Food, or Microorganisms Causing Human Respiratory Diseases

Disease	Brief Description of Disease	Organisms Causing Disease	How Disease Is Transmitted from Person to Person	Other Common Routes of Transmission (Indirectly from Person to Person)	How Pathogens Are Transmitted to Food and Then Customers	How to Prevent Transmission of Disease in the Foodservice Department
Common Cold	Sniffles, cough, sinus infection, bronchial infection.	Viruses and bacteria	Germ-laden droplets expelled through sneezing or coughing; discharges from nose and throat; touching with hands soiled by sneezing or coughing or by handling soiled handkerchiefs.	Articles touched, coughed at or sneezed at by a sick person; these articles include food, work surfaces, and utensils used in food preparation and serving.	Through sick or carrier food handlers who prepare and serve food or who handle eating utensils, pots and pans. Through sick or carrier customers who contaminate displayed food or whose eating utensils are not properly sanitized.	Food handlers ill with a cold must not prepare and serve food or handle pots, pans, or eating utensils. Food handlers must cover the mouth when coughing or sneezing, and then wash hands thoroughly. Displayed food must not be touched by customers or exposed to their sneezes or cough droplets. Effective dishwashing must always be practiced.

(continued)

Table I. (Continued)

Disease	Brief Description of Disease	Organisms Causing Disease	How Disease Is Transmitted from Person to Person	Other Common Routes of Transmission (Indirectly from Person to Person)	How Pathogens Are Transmitted to Food and Then Customers	How to Prevent Transmission of Disease in the Foodservice Department
Septic Sore Throat	Sore throat, fever, body aches, swelling of lymph glands. A person recovered from illness may become a carrier.	Streptococci (hemolytic bacteria).	Most common: germ-laden droplets expelled from throat; soiled hands.	Soiled handkerchiefs and other articles soiled with discharges from the mouth. Food and dishes contaminated with sputum.	Through sick or carrier food handlers who contaminate food and other kitchen objects with discharge from nose and throat. Contaminated eating utensils.	Sick food handlers or carriers must not prepare or serve food nor handle food preparation equipment or eating utensils. Before the patient returns to work, a check should be made by physician to make sure that the patient has not become a carrier. Sanitary dishwashing must always be practiced.
Scarlet Fever (often epidemic)	A highly contagious communicable disease that starts with sore throat, fever, and a skin rash. Carriers are possible.	Streptococci (hemolytic bacteria).	Contact with diseased persons and those recuperating. Strict isolation is required in many states for 3 weeks.	All articles that have had contact with disease and convalescent persons, including food and clothing.	Through sick or convalescing food handlers, and "healthy" carriers. Contaminated eating utensils.	(See Septic Sore Throat)

Infection of Tonsils and Adenoids	Fever, sore larynx, cough, sometimes inflammation of eyes, bronchia, and lungs	Viruses	Coughing	(See Septic Sore Throat)	(See Septic Sore Throat)	(See Septic Sore Throat)
Pneumonia (bacterial)	Inflammation of the lungs	*Diplococcus pneumoniae*, a bacterium	Droplet infection from a diseased person or carrier.	Hands contaminated with saliva; bacteria can be transmitted through food contaminated through coughing or from contact with contaminated hands.	Through sick or carrier food handlers who contaminate food and other kitchen objects with discharge from nose and throat. Contaminated eating utensils.	Sick food handlers or carriers must not prepare or serve food nor handle food preparation equipment or eating utensils. Before the patient returns to work, a check should be made by physician to make sure that the patient has not become a carrier.
Tuberculosis of the Lungs (pulmonary)	Chest pains, coughing, fever, fatigue, loss of weight. Through X ray of chest, pulmonary tuberculosis can be detected. Bacteria are discharged for a long time by the recuperating patient.	*Mycobacterium tuberculosis*, a bacterium	Most often by droplet infection (coughing); or by fingers soiled with saliva; or by sputum discharged from lungs; or by feces and urine.	Articles soiled with discharge from mouth; or soiled with feces or urine.	Through a sick food handler. Contaminated eating utensils.	Food handlers with tuberculosis must not be allowed to work in a food-service establishment.

(continued)

Table I. (Continued)

Disease	Brief Description of Disease	Organisms Causing Disease	How Disease Is Transmitted from Person to Person	Other Common Routes of Transmission (Indirectly from Person to Person)	How Pathogens Are Transmitted to Food and Then Customers	How to Prevent Transmission of Disease in the Foodservice Department
Diphtheria (rare today because of immunization)	At first sore throat and fever. A choking membranelike structure forms in the throat which may extend deeply and suffocate the patient; a powerful toxin poisons the victim's vital organs. Carriers are common.	*Corynebacterium diphtheriae*, a bacterium	Discharge from mouth, throat, and nose.	Articles soiled with discharge from mouth, throat, and nose; contaminated food and eating utensils.	Through sick and carrier food handlers who prepare food or handle eating utensils, pots and pans. Contaminated eating utensils.	Sick food handlers or carriers must not prepare and serve food or handle pots, pans, utensils, dishes, glasses, and flatware. Before the person returns to work, a physician should certify that he or she is not a carrier.
Influenza (epidemic); also called "Flu"	Fever, head cold, pharyngitis and other respiratory infections; also, headache, sore throat, muscle pain, weakness.	Viruses	Direct contact; also droplet infection.	Articles soiled with discharge from mouth and nose of sick person; contaminated food and eating utensils.	Through sick food handlers who prepare food or handle eating utensils, pots and pans. Contaminated eating utensils.	Sick food handlers must not prepare and serve food while ill or handle pots, pans, utensils, dishes, glasses, or flatware.

Table II. Transmission, by Way of Food, of Microorganisms Causing Human Intestinal Diseases

Disease	Brief Description Of Disease	Organisms Causing Disease	How Disease Is Transmitted Directly from Person to Person	Other Common Routes of Transmission	How Pathogens Are Transmitted to Food and Then Customers	How Disease Transmission Can Be Prevented in the Foodservice Department
Typhoid Fever	Infection of intestinal tract, continued fever, headache, abdominal pain. Carriers are common.	*Salmonella typhi,* a bacterium	By touch, from hands contaminated with the excreta of a diseased person or a carrier.	Through: 1. Food handled by a person who is acutely ill or a carrier; this may happen in a kitchen or dairy. 2. Sewage-polluted water. 3. Shellfish harvested from sewage-polluted water and eaten raw. 4. Food visited by flies and other vermin.	Usually through hands of food handler who is or has been ill with typhoid, and is a "healthy" carrier; through drinking sewage-polluted water or eating sewage-polluted food; through contact with vermin such as rodents and flies.	Persons who are ill or carriers must not be allowed in the food preparation and service area and must not touch pots, pans, dishes, glasses, or flatware. Use pure water. Use nonleaky sewer pipes and proper sewage disposal. Use pasteurized milk and other dairy products. Use shellfish from unpolluted water. Practice fly and rodent control.

(continued)

Table II. (continued)

Disease	Brief Description Of Disease	Organisms Causing Disease	How Disease Is Transmitted Directly from Person to Person	Other Common Routes of Transmission	How Pathogens Are Transmitted to Food and Then Customers	How Disease Transmission Can Be Prevented in the Foodservice Department
Bacillary Dysentery (Shigellosis)	A very common illness; diarrhea caused by bacteria (there is another type caused by amebae). Carriers are common.	Several species of *Shigella*, a bacterium	By touch and from clothing soiled with excreta of a person who is ill, or a carrier.	Contaminated food; contaminated water; contaminated flies.	Hands of food handler, ill or carrier; leaky sewer pipes.	Persons who are ill or carriers must not be allowed in the food preparation and service area, and must not touch pots, pans, dishes, glasses, or flatware. Use pure water. Use nonleaky sewer pipes and proper sewage disposal. Practice fly and rodent control.
Amebic Dysentery	Symptoms are more or less mild and range from mere abdominal discomfort to slight diarrhea alternating with constipation to severe diarrhea. Carriers are common. A person is likely to remain a carrier for life.	*Entameba histolytica*, a protozoan	By touch from hands soiled with excreta of contaminated person and by swallowing these contaminants. Many carriers are symptomless.	Contaminated food and water; contaminated flies and rodents, especially rats.	Food and drink handled by diseased persons, convalescents and carriers; sewage-polluted water supply (cross-connections in plumbing); contaminated flies and rodents that feed on unprotected food items.	Persons who are ill, or carriers, must not handle food or utensils used in the preparation and service of food. Use pure water. Control vermin. Use proper sewage control.

Disease	Cause	Symptoms	How Spread	Sources	Common Sources	Prevention
Asiatic Cholera	*Vibrio comma*, a bacterium	Rare in the Western world. Intestinal pain and diarrhea, vomiting. Usually fatal. Healthy carriers are not common.	By touch from hands soiled with excreta of contaminated person and by swallowing the contaminants.	Contaminated food and water; contaminated flies and rodents, especially rats.	Food and drink handled by ill persons; food items exposed to flies and rodents.	See Amebic Dysentery.
Infectious Hepatitis	Viruses	Infection of the liver. First, fever, nausea, weakness, abdominal pain; then jaundice. Carriers are common.	By touch from hands soiled with excreta of contaminated person and by swallowing the contaminants.	Sewage-contaminated drinking water; sewage-polluted shellfish; food contaminated with excreta of food handler; food contaminated from leaky sewer pipes or other faulty plumbing; flies.	Drinking water supply contaminated through faulty plumbing; use of shellfish harvested from sewage-polluted waters; preparation of food by ill or carrier food handler; faulty plumbing; poor fly control.	Use uncontaminated water supply; purchase and use shellfish from approved sources. Keep toilets in excellent order. Prevent persons ill with hepatitis and carriers from working in food preparation and service. Practice fly control. Keep plumbing in excellent order. Prevent sewage from dripping on food or food-preparation surfaces.

Table III. Transmission, by Way of Food, of Microorganisms Causing Diseases in Animals and Humans

Disease	Brief Description of Disease in Humans	Organisms Causing Disease	How Disease Is Usually Transmitted	How Pathogens Are Transmitted to Food and Customers	How to Prevent Transmission
Brucellosis A disease of cows; can cause disease in humans. (In the human, brucellosis is also called Malta fever or undulant fever.)	Causes irregular fever, sweating, headache, sore throat, pain in joints and muscles.	*Brucella abortus*, a bacterium	Consumption of unpasteurized milk from infected cows.	Use of unpasteurized milk from infected cows.	Use only pasteurized milk.
Tularemia A disease of wild rabbits, squirrels, rodents, and birds. May be transmitted to humans.	Causes fever, headache, vomiting, chills, abcess at point of infection.	*Pasteurella tularensis*, a bacterium	Skinning or otherwise handling diseased animals; bite of a tick that had inhabited an infected animal; eating improperly cooked wild rabbit and squirrel.	Use of diseased meat from rabbits, or squirrels, especially when undercooked.	Do not prepare wild rabbit or squirrel for public consumption.

Trichinosis		*Trichinella spiralis*, a nematode		
A disease of hogs, rabbits, and bears caused by a nematode worm. Disease is spread by feeding hogs infested, uncooked garbage. Other host animals whose meat might serve as human food are rabbits and bears. Trichinae cause disease in humans when infested, undercooked meat is eaten.	First symptoms resemble food poisoning (nausea, vomiting, diarrhea, colic). Later-on, muscular soreness and swelling. Death may ensue in severe infestation.	Consumption of undercooked, infested meat and sausage. In the U.S.A. incidences of trichinosis in hogs are declining steadily; so are incidences of trichinosis in humans.	The larvae of the trichinae are already in the infested meat when cooked; when such meat is undercooked and trichinae are eaten, these are released into intestinal tract of the human. May propagate in the human and finally settle in the muscle tissues.	Purchase and serve inspected pork and pork products. Serve all pork (including sausage) well done. Roasts should reach a temperature of at least 165 to 170 F (74 to 77 C) in the center of the cut. Safe end point is reached when the meat has turned from pink to gray. Cook all small cuts and sausage to the gray stage.

Table IV. Botulism

Illness and Incidence	Brief Description of Illness	Cause and Main Source of Organisms Involved	Growth Requirements, Heat Resistance of Organism	Food Items Frequently Accused of Causing Illness	Circumstances Often Associated with Outbreaks	Principles of Control
Botulinum Food Poisoning (Botulism) Incidence: Infrequent	Fatality high, 65% of cases. Symptoms appear approximately 12 to 36 hours following consumption of the contaminated food; digestive disturbances; nausea; vomiting; diarrhea; dizziness; dryness of throat; double vision; difficulty in swallowing and speaking, followed by increasing	Toxin of *Clostridium botulinum,* a spore former. Types A, B and E of importance in U.S. Source of organism: soil, for Type A and B; for Type E, sediment of lakes, rivers, and oceans.	Growth and toxin production without air (anaerobically) and at a wide range of temperatures; for growth temperatures: Type A and B: 60 to 118 F (15.5 to 48 C); Type E, 38 to 113 F (3.3 to 45 C). Heat resistance: *spores* are killed when foods are processed by	Canned foods of low acidity that have been underprocessed, usually home-canned. Damaged, leaky, rusty cans; cans or jars with broken seals. Contents may or may not have a spoiled appearance. Smoked products that were under- processed (not heated	Eating, or merely tasting, suspect items.	*Canned foods:* Use commercially canned foods when preparing food for the public. Refuse donations of home-canned goods. Store canned foods in cool room. Practice quick turnover of canned foods. Do not keep cans long. A year is the longest acceptable period for storing canned foods. Inspect canned goods

paralysis of involuntary muscles. Victim dies of inability to breathe.

proper methods of heating. Toxin is inactivated by boiling for 15 minutes.

Toxins in canned foods can be destroyed by boiling the food for 15 minutes.

sufficiently to destroy the spores) and then held above 38 F (3.3 C).

closely before opening for bulging, damaged, rusty, and leaky cans. Contents are suspect if they spurt out on opening the can, look bubbly, or are off-odor or off-color. Toxin, if present, is destroyed by boiling the food for 15 minutes.

DO NOT EVER TASTE SUSPECT FOOD: Boil for 15 minutes and DISCARD.

Smoked fish:
Keep smoked fish frozen or under strict refrigeration, below 38 F (3.3 C). Use smoked fish within a few days of refrigerator storage.

Table V. Staphylococcus Food Poisoning

Illness and Incidence	Brief Description of Illness	Cause and Main Source of Organisms Involved	Growth Requirements, Heat Resistance of Organism	Food Items Frequently Accused of Causing Illness	Circumstances Often Associated with Outbreaks	Principles of Control
Staphylococcus Food Poisoning Incidence: very frequent	Symptoms usually appear 2 to 3 hours (½ to 6 hours) after consuming the contaminated item. They are nausea, vomiting, diarrhea, abdominal cramps, all more or less violent; headaches, sweating, weakness. Fatal cases are rare, but may occur when victims	Toxin of *Staphylococcus aureus* *Source:* The healthy human body, especially the respiratory tract. Particularly rich in food-poisoning staph is the infected respiratory tract; infected	*Growth and toxin production:* Growth occurs between 39 and 114 F (4 and 46 C). Excellent growth between 70 and 98.7 F (21 and 37 C) *Heat resistance:* Heating to 165 F (74 C) kills most of the bacteria themselves, but	Protein foods, especially those that have undergone much handling (as in chopping, slicing, shaping, or deboning); foods that have been exposed to lukewarm temperatures for more than a couple of hours; and foods	Food handler who was ill with respiratory disease; skin, eye or ear infection. Healthy food handler who failed to practice good personal hygiene, including frequent, thorough washing of hands.	Persons suffering from colds, influenza, sore throat, sinus infections, and pus-containing sores should not handle food. All food handlers must be clean. They must not cough or sneeze onto food and must wash their hands after sneezing or coughing.

Hands must be washed frequently during preparation of food: after touching likely sources of staph (the person's own body, and other contaminated items such as hand-kerchiefs).

Utensils rather than hands should be used to pick up and manipulate food.

Food should be kept hot, 140 F (60 C) or higher; or cold, 45 F (7.2 C) or lower.

Warm food should be cooled rapidly to safe temperatures.

Food that was exposed to contaminated hands or equipment.

Food that was exposed to warm temperatures for long hours.

Leftovers were mishandled.

that are insufficiently refrigerated because of large bulk or high refrigerator temperature; leftovers in general. *Examples:* cream-filled pastries, poultry and meat pies, sandwiches with egg or meat fillings, casseroles and croquettes made with leftovers, chicken salad, potato salad, creamed items, and many others too numerous to mention.

not the toxins; *toxins* are resistant to heating. Staph are salt-loving bacteria.

cuts, burns and other wounds; infected ears and eyes; and pimples and boils.

are very young, very old, feeble, or ill.

Table VI. Salmonella Food Infection (Salmonellosis)

Illness and Incidence	Brief Description of Illness	Cause and Main Source of Organisms Involved	Growth Requirements, Heat Resistance of Organism	Food Items Frequently Accused of Causing Illness	Circumstances Often Associated with Outbreaks	Principles of Control
Salmonella Food Infection (Salmonellosis) Incidence: very frequent	Symptoms appear usually 12 to 24 hours after consumption of the contaminated item. They include nausea, vomiting, diarrhea, abdominal cramps, all more or less violent; headaches, chills, weakness, sometimes fever. Death is not uncommon if the afflicted are very young or aged or infirm.	Infection caused by consuming food containing salmonellae in large numbers. Many species and strains of *Salmonella* are capable of causing salmonellosis. *Source of organism:* the intestinal tract of humans and animals and their waste products such as feces and urine, sewage, manure, and droppings; and everything contaminated with human or animal excrements which includes human	No toxin is released into food during multiplication; the bacteria set up infection in the gastrointestinal tract. Growth occurs in food at temperatures between 44 and 115 F (6.7 and 46.6 C). Excellent growth is possible between 70 and 98.7 F (21 and 37 C). *Heat resistance* Heating to 165 F (74 C) kills the majority of salmonellae.	Protein foods are excellent food for these bacteria. Meats, poultry, and cracked eggs are likely to be already contaminated when received and are doubly hazardous. Food items that have very frequently caused *Salmonella* food infections are turkey, stuffing and gravy; meat and poultry pies; cracked, undercooked eggs; chicken and turkey salads; meringue pies; and others too numerous to mention.	Food handler was ill with salmonellosis or who just recovering or who are carriers, must not handle food. Food handler did not wash, or superficially washed, hands after a visit to the toilet. Food handler did not wash hands after handling raw meat and poultry. Food handler did not thoroughly clean and sanitize chopping blocks or other surfaces and equipment after using them for raw meat and poultry. Rodents and flies had access to kitchen. Seafood from polluted water was used.	Persons suffering from salmonellosis or who are just recovering or who are carriers, must not handle food. All food handlers must thoroughly wash their hands after a visit to the toilet. Before starting work, the worker must wash hands, and frequent hand washing must be practiced during preparation: after touching anything other than food, after touching food likely to be contaminated with salmonellae—meat, poultry, egg shells—and before touching other foods, especially cooked foods.

hands, sewage-polluted water, and seafood; meat and poultry contaminated with animal excrement at slaughter; food preparation surfaces and equipment contaminated by contact with raw meat and poultry; food or surfaces contaminated by rodents and insects or leaking sewage (faulty plumbing).

Large stuffed turkey was prepared and undercooked in center; contaminated giblets were added to gravy. Cracked eggs were used. Cooked food was exposed to warm temperatures. Leftovers were mishandled.

Rodents and insects must be controlled.

Food-preparation surfaces, equipment, and utensils must be kept in a sanitary condition. It is advised that *separate cutting surfaces be used for raw and cooked items.*

Food should not be held at lukewarm temperatures. Keep food hot, 140 F (60 C) or above; or cold, 45 F (7.2 C) or below.

Warm food should be cooled rapidly to safe temperatures.

Table VII. Perfringens Food Poisoning

Illness and Incidence	Brief Description of Illness	Cause and Main Surce of Organisms Involved	Growth Requirements, Heat Resistance of Organism	Food Items Frequently Accused of Causing Illness	Circumstances Often Associated with Outbreaks	Principles of Control
Perfringens Food Poisoning Incidence: less frequent than food-borne illnesses caused by staph and salmonellae	Symptoms usually appear 8 to 22 hours after consuming the contaminated item. They are similar to those caused by staph and salmonellae, but milder.	*Clostridium perfringens* *Source* of organism: Intestinal tract of humans and animals, soil, dust, soiled equipment and utensils. Meats have good chance to become	Growth occurs between 59 and 122 F (15 and 50 C). Between 109 and 116 F (43 and 47 C) growth is excellent. *Heat resistance:* Some strains are very resistant to heat. We cannot rely on killing the spores by boiling,	Meat and gravy are involved most frequently, especially beef and veal. Rolled roasts are apt to be contaminated on the inside, and are particularly hazardous when cooked to the rare stage.	Meat was cooked the previous day or several hours in advance, giving the bacteria a chance to multiply. Menu items were heavily contaminated with perfringens bacteria adhering to hands and soiled equip-	Since the organism is everywhere, its presence in the kitchen—in dust, on equipment, on work tables, and on and in food—must be anticipated. Hands must be washed after raw meat has been handled. Work surfaces and

food preparation equipment and utensils must be maintained in sanitary condition.

C. perfringens is a spore former. Heating used in ordinary cooking cannot be relied on for killing the cells; therefore food must be kept out of the danger zone of bacterial growth as much as is possible. Keep food either hot, 140 F (60 C) or above; or cold (45 F (7.2 C) or below. Warm food should be cooled rapidly to safe temperatures.

ment previously used for raw meat. Contaminated items were held neither hot enough nor cold enough.

contaminated at time of slaughter. However, many other foods have also been found to be contaminated with *C. perfringens*.

and must prevent organism from multiplying.

Sample Foodservice Establishment Inspection Report

DEPARTMENT OF HEALTH, EDUCATION AND WELFARE
PUBLIC HEALTH SERVICE - FOOD AND DRUG ADMINISTRATION
FOOD SERVICE ESTABLISHMENT INSPECTION REPORT

Based on an inspection this day, the items circled below identify the violation in operations or Facilities which must be corrected by the next routine inspection or such shorter period of time as may be specified in writing by the regulatory authority. Failure to comply with any time limits for corrections specified in this notice may result in cessation of your Food Service operations.

OWNER NAME

ESTABLISHMENT NAME

ADDRESS

ZIP CODE

ESTABLISHMENT I.D.				CENSUS TRACT	SANIT. CODE	DATE			INSPECT. TIME (Min.)	PURPOSE	29
COUNTY	DISTRICT	TYPE	EST. NO.			YR.	MO.	DAY			
1 2 3	4 5 6	7 8 9	10 11 12 13	14 15 16	17 18 19	20 21 22	23 24	25 26	27 28	Regular.................1 Complaint 3	
										Follow-up ....2 Investigation.................4	
										Other.................5	

ITEM	WT.	COL.	ITEM	WT.	COL.	ITEM	WT.	COL.
FOOD			18 PRE-FLUSHED, SCRAPED, SOAKED	1	47	**GARBAGE AND REFUSE DISPOSAL**		
*01 SOURCE; SOUND CONDITION, NO SPOILAGE	5	30	19 WASH, RINSE WATER: CLEAN, PROPER TEMPERATURE	2	48	33 CONTAINERS OR RECEPTACLES, COVERED; ADEQUATE NUMBER, INSECT/RODENT PROOF, FREQUENCY, CLEAN	2	62
02 ORIGINAL CONTAINER; PROPERLY LABELED	1	31	*20 SANITIZATION RINSE: CLEAN, TEMPERATURE, CONCENTRATION, EXPOSURE TIME, EQUIPMENT, UTENSILS SANITIZED	4	49	34 OUTSIDE STORAGE AREA ENCLOSURES PROPERLY CONSTRUCTED, CLEAN; CONTROLLED INCINERATION	1	63
FOOD PROTECTION			21 WIPING CLOTHS: CLEAN, STORED, RESTRICTED	1	50	**INSECT, RODENT, ANIMAL CONTROL**		
*03 POTENTIALLY HAZARDOUS FOOD MEETS TEMPERATURE REQUIREMENTS DURING STORAGE, PREPARATION, DISPLAY, SERVICE, TRANSPORTATION	5	32	22 FOOD CONTACT SURFACES OF EQUIPMENT AND UTENSILS CLEAN, FREE OF ABRASIVES, DETERGENTS	2	51	*35 PRESENCE OF INSECT/RODENTS – OUTER OPENINGS PROTECTED, NO BIRDS, TURTLES, OTHER ANIMALS	4	64
*04 FACILITIES TO MAINTAIN PRODUCT TEMPERATURE	4	33	23 NON-FOOD CONTACT SURFACES OF EQUIPMENT AND UTENSILS CLEAN	1	52	**FLOORS, WALLS AND CEILINGS**		
05 THERMOMETERS PROVIDED AND CONSPICUOUS	1	34	24 STORAGE, HANDLING OF CLEAN EQUIPMENT/UTENSILS	1	53	36 FLOORS: CONSTRUCTED, DRAINED, CLEAN, GOOD REPAIR, COVERING INSTALLATION, DUSTLESS CLEANING METHODS	1	65
06 POTENTIALLY HAZARDOUS FOOD PROPERLY THAWED	2	35	25 SINGLE-SERVICE ARTICLES, STORAGE, DISPENSING, USED	1	54			
*07 UNWRAPPED AND POTENTIALLY HAZARDOUS FOOD NOT RE-SERVED	4	36	26 NO RE-USE OF SINGLE SERVICE ARTICLES	2	55	37 WALLS, CEILING, ATTACHED EQUIPMENT: CONSTRUCTED, GOOD REPAIR, CLEAN SURFACES, DUSTLESS CLEANING METHODS	1	66
08 FOOD PROTECTION DURING STORAGE, PREPARATION, DISPLAY, SERVICE, TRANSPORTATION	2	37	**WATER**					
09 HANDLING OF FOOD (ICE) MINIMIZED	2	38	*27 WATER SOURCE, SAFE: HOT AND COLD UNDER PRESSURE	5	56	**LIGHTING**		
10 IN USE, FOOD (ICE) DISPENSING UTENSILS PROPERLY STORED	1	39	**SEWAGE**			38 LIGHTING PROVIDED AS REQUIRED, FIXTURES SHIELDED	1	67
PERSONNEL			*28 SEWAGE AND WASTE WATER DISPOSAL	4	57	**VENTILATION**		
*11 PERSONNEL WITH INFECTIONS RESTRICTED	5	40	**PLUMBING**			39 ROOMS AND EQUIPMENT VENTED AS REQUIRED	1	68
*12 HANDS WASHED AND CLEAN, GOOD HYGIENIC PRACTICES	5	41	29 INSTALLED, MAINTAINED	1	58	**DRESSING ROOMS**		
13 CLEAN CLOTHES, HAIR RESTRAINTS	1	42	*30 CROSS-CONNECTION, BACK SIPHONAGE, BACKFLOW	5	59	40 ROOMS CLEAN, LOCKERS PROVIDED. FACILITIES CLEAN, LOCATED, USED	1	69
FOOD EQUIPMENT AND UTENSILS			**TOILET AND HANDWASHING FACILITIES**			**OTHER OPERATIONS**		
14 FOOD (ICE) CONTACT SURFACES: DESIGNED, CONSTRUCTED, MAINTAINED, INSTALLED, LOCATED	2	43	*31 NUMBER, CONVENIENT, ACCESSIBLE, DESIGNED, INSTALLED	4	60	*41 NECESSARY TOXIC ITEMS PROPERLY STORED, LABELED, USED	5	70
15 NON-FOOD CONTACT SURFACES: DESIGNED, CONSTRUCTED, MAINTAINED, INSTALLED, LOCATED	1	44	32 TOILET ROOMS ENCLOSED, SELF-CLOSING DOORS, FIXTURES, GOOD REPAIR, CLEAN HAND CLEANSER, SANITARY TOWELS/TISSUE/HAND-DRYING DEVICES PROVIDED, PROPER WASTE RECEPTACLES	2	61	42 PREMISES MAINTAINED, FREE OF LITTER, UNNECESSARY ARTICLES, CLEANING MAINTENANCE EQUIPMENT PROPERLY STORED, AUTHORIZED PERSONNEL	1	71
16 DISHWASHING FACILITIES: DESIGNED, CONSTRUCTED, MAINTAINED, INSTALLED, LOCATED, OPERATED	2	45				43 COMPLETE SEPARATION FROM LIVING/SLEEPING QUARTERS. LAUNDRY	1	72
17 ACCURATE THERMOMETERS, CHEMICAL TEST KITS PROVIDED, GAUGE COCK (¼" IPS VALVE)	1	46				44 CLEAN, SOILED LINEN PROPERLY STORED	1	73

FOLLOW-UP   74
YES 1
NO 2

RATING SCORE
("100" Less Weight of Items Violated)   75 76 77

* CRITICAL ITEMS REQUIRING IMMEDIATE ACTION

RECEIVED BY (Name and Title)

INSPECTED BY (Name and Number and Title)

Poster: Job Safety and Health Protection (OSHA)

job safety and health protection

The Occupational Safety and Health Act of 1970 provides job safety and health protection for workers through the promotion of safe and healthful working conditions throughout the Nation. Requirements of the Act include the following:

Employers: Each employer shall furnish to each of his employees employment and a place of employment free from recognized hazards that are causing or are likely to cause death or serious harm to his employees; and shall comply with occupational safety and health standards issued under the Act.

Employees: Each employee shall comply with all occupational safety and health standards, rules, regulations and orders issued under the Act that apply to his own actions and conduct on the job.

The Occupational Safety and Health Administration (OSHA) of the Department of Labor has the primary responsibility for administering the Act. OSHA issues occupational safety and health standards, and its Compliance Safety and Health Officers conduct jobsite inspections to ensure compliance with the Act.

Inspection: The Act requires that a representative of the employer and a representative authorized by the employees be given an opportunity to accompany the OSHA inspector for the purpose of aiding the inspection.

Where there is no authorized employee representative, the OSHA Compliance Officer must consult with a reasonable number of employees concerning safety and health conditions in the workplace.

Complaint: Employees or their representatives have the right to file a complaint with the nearest OSHA office requesting an inspection if they believe unsafe or unhealthful conditions exist in their workplace. OSHA will withhold, on request, names of employees complaining.

The Act provides that employees may not be discharged or discriminated against in any way for filing safety and health complaints or otherwise exercising their rights under the Act.

An employee who believes he has been discriminated against may file a complaint with the nearest OSHA office within 30 days of the alleged discrimination.

Citation: If upon inspection OSHA believes an employer has violated the Act, a citation alleging such violations will be issued to the employer. Each citation will specify a time period within which the alleged violation must be corrected.

The OSHA citation must be prominently displayed at or near the place of alleged violation for three days, or until it is corrected, whichever is later, to warn employees of dangers that may exist there.

Proposed Penalty: The Act provides for mandatory penalties against employers of up to $1,000 for each serious violation and for optional penalties of up to $1,000 for each nonserious violation. Penalties of up to $1,000 per day may be proposed for failure to correct violations within the proposed time period. Also, any employer who willfully or repeatedly violates the Act may be assessed penalties of up to $10,000 for each such violation.

Criminal penalties are also provided for in the Act. Any willful violation resulting in death of an employee, upon conviction, is punishable by a fine of not more than $10,000 or by imprisonment for not more that six months, or by both. Conviction of an employer after a first conviction doubles these maximum penalties.

Voluntary Activity: While providing penalties for violations, the Act also encourages efforts by labor and management, before an OSHA inspection, to reduce injuries and illnesses arising out of employment.

The Department of Labor encourages employers and employees to reduce workplace hazards voluntarily and to develop and improve safety and health programs in all workplaces and industries.

Such cooperative action would initially focus on the identification and elimination of hazards that could cause death, injury, or illness to employees and supervisors. There are many public and private organizations that can provide information and assistance in this effort, if requested.

More Information: Additional information and copies of the Act, specific OSHA safety and health standards, and other applicable regulations may be obtained from your employer or from the nearest OSHA Regional Office in the following locations:

Atlanta, Georgia
Boston, Massachusetts
Chicago, Illinois
Dallas, Texas
Denver, Colorado
Kansas City, Missouri
New York, New York
Philadelphia, Pennsylvania
San Francisco, California
Seattle, Washington

Telephone numbers for these offices, and additional Area Office locations, are listed in the telephone directory under the United States Department of Labor in the United States Government listing.

Washington, D.C.
1977
OSHA 2203

Ray Marshall

Ray Marshall
Secretary of Labor

U. S. Department of Labor
Occupational Safety and Health Administration

INDEX